THE THEORY OF

Island
Biogeography

PRINCETON LANDMARKS IN BIOLOGY

The Theory of Island Biogeography
by Robert H. MacArthur and Edward O. Wilson
With a new preface by Edward O. Wilson

THE THEORY OF
Island
Biogeography

ROBERT H. MACARTHUR AND

EDWARD O. WILSON

With a new preface by Edward O. Wilson

PRINCETON UNIVERSITY PRESS

PRINCETON AND OXFORD

Thirteenth printing, and first Princeton Landmarks in Biology edition,
with a new preface by Edward O. Wilson, 2001

Library of Congress Cataloging-in-Publication Data

MacArthur, Robert, Robert H.
    The theory of island biography / Robert H. MacArthur and
Edward O. Wilson ; with a new preface by Edward O. Wilson.
        p. cm. — (Princeton landmarks in biology)
    Originally published: Princeton, N.J. : Princeton University Press,
1967.
    Includes bibliographical references and index (p.   ).
    ISBN 0-691-08836-5 (pbk. : alk. paper)
    1. Biogeography. 2. Island ecology. I. Wilson, Edward
Osborne, 1929– II. Title.

QH85 M3 2001
578.75'2—dc21   00-051495

Printed on acid-free paper. ∞

www.pup.princeton.edu

Printed in the United States of America

20 19 18 17 16 15

Contents

Preface to the Princeton
Landmarks in Biology Edition

Robert Helmer MacArthur's death from cancer in 1972, at the very early age of 42, deprived ecology of a creative genius who, with less than two decades of research, had already left a permanent impression on his discipline. He was a gifted mathematician, which in the notably nonanalytical milieu of the time was a key to his success. But even more importantly, he was a dedicated naturalist with a deep love and understanding of birds, his favorite group of organisms. He combined the logic and imaginative process of mathematics with the fingertip feel of ornithology to create simplifying models of complex phenomena in evolutionary and community ecology, and thereby set the tone for an entire generation's effort in theoretical studies.

It was my privilege to collaborate with Robert (he preferred not to be called Bob) on island biogeography, arguably the piece of work for which he is best and most favorably remembered today. The success of our effort was due to *The Theory of Island Biogeography*, published in 1967. In this summary work, we spelled out the likely parameters of the assembly of discrete biotic communities, and then, not satisfied, struck out from this base to explore other, related phenomena, such as demography and competition.

When our friendship began in 1960, I brought to our discussions two of the main elements of island biogeography. The first was a detailed knowledge of animal distribution and ecology and especially of the ants of Melanesia, on which I had recently conducted field and museum studies. The second was the theory, developed earlier by William Diller Matthew, Philip J. Darlington, and George G. Simpson, of faunistic replacement and balance in the major groups of Cenozoic land vertebrates. These authors had conceived of a rough equilibrium in continental faunas at the taxonomic level of families. Matthew and Darlington, in particular, envisaged a cyclical pattern of newly dominant groups replacing older ones, only to retreat themselves before fresh competitors across evolutionary time. Subsequently, I worked out similar patterns at the finer, species level in the ants of New Guinea and neighboring archipelagoes. In the process, I documented the logarithmic relation of species to island area in ants and a few other animal and plant groups. All this I brought to MacArthur's attention and suggested that we might create a more rigorous theory of biogeography than had hitherto been possible. He responded by devising the famous crossing curve of species immigration and extinction. We were then off and running in broader exchanges of ideas and data on the implications of biotic equilibrium for ecology. In 1963 we published a bare-bones article, "An Equilibrium Theory of Insular Zoogeography" (*Evolution* 17:363–387), and followed it in 1967 with *The Theory of Island Biogeography*. The article was scarcely noticed, but the book was an instant critical and—at least by university press standards—commercial success.

The Theory of Island Biogeography has exerted an important influence on both biogeography and ecology. True, after more than three decades, it has been largely replaced by a generation of far more detailed and sophisticated studies. Yet I believe that its basic structure remains sound, and the content still serves as a good introduction to the subject. Furthermore, in a way that MacArthur and I failed to appreciate, the book has had a major impact on conservation biology. This young discipline grew significantly in the 1970s and came into flower in the 1980s and 1990s. Because *The Theory of Island Biogeography* deals centrally with habitat fragmentation, hence insularization, the creation of biotic communities, and species extinction, its relevance to conservation biology was immediately clear.

The flaws of the book lie in its oversimplification and incompleteness, which are endemic to most early efforts at theory and synthesis. Large numbers of experiments and field data on many biotas supported the hypothesis of species equilibrium, but many others did not. In some cases the variance was explained—for example, by frequent environmental disturbance and chronic disequilibria—but often no clear cause was adduced, and in the latter circumstance the basic theory found little use except perhaps to stimulate further study. Also, MacArthur and I had been satisfied to account for the effect of area on equilibrial species numbers as an outcome of varying population size and fluctuation. Thus, small islands, supporting small populations, are more prone to lose species than large ones, and the effect is exacerbated when the amplitude of population fluctuation is increased. Later, others were quick to point out that population size is far from the whole story. The area effect owes a lot to the happenstance of physical geography. In particular, large islands have more variable topography, soil types, and other determining features of vegetation and microclimate, which in turn affect colonization and extinction rates.

Thus, island biogeography has evolved into a subject far more enmeshed in the particularities of natural history. Departing from the early models of species equilibrium, it has also engaged most other disciplines of biology, including population genetics, life-cycle studies, ethology, and ecosystems studies. To a considerable degree, it has been dissipated into them. I call that progress and would have it no other way, and I am certain Robert MacArthur, if he were here, would agree.

Edward O. Wilson
Harvard University
November 1999

Preface

This book had its origin when, about five years ago, an ecologist (MacArthur) and a taxonomist and zoogeographer (Wilson) began a dialogue about common interests in biogeography. The ideas and the language of the two specialties seemed initially so different as to cast doubt on the usefulness of the endeavor. But we had faith in the ultimate unity of population biology, and this book is the result. Now we both call ourselves biogeographers and are unable to see any real distinction between biogeography and ecology.

A great deal of faith in the feasibility of a general theory is still required. We do not seriously believe that the particular formulations advanced in the chapters to follow will fit for very long the exacting results of future empirical investigation. We hope instead that they will contribute to the stimulation of new forms of theoretical and empirical studies, which will lead in turn to a stronger general theory and, as R. A. Fisher once put it, "a tradition of mathematical work devoted to biological problems, comparable to the researches upon which a mathematical physicist can draw in the resolution of special difficulties." Already some strains have appeared in the structure. These have been discussed frankly, if not always satisfactorily, in the text.

We owe the strains, as well as many improvements, to colleagues who read the entire first draft. We are very grateful to John T. Bonner, William L. Brown, Jr., Walter Elsasser, Carl Gans, Henry Horn, Robert F. Inger, E. G. Leigh, Richard Levins, Daniel A. Livingstone, Monte Lloyd, Thomas Schoener, and Daniel Simberloff for this favor. We are also indebted to William H. Bossert, Philip J. Darlington, Bassett Maguire, Ernst Mayr, and Lawrence B. Slobodkin for critically reading selected portions of the manuscript; and to J. Bruce Falls, Kenneth Crowell,

xi

Bassett Maguire, Ruth Patrick, and Bernice G. Schubert for adding new materials. A preliminary draft of the book was used as a text in graduate seminars at Harvard University and Princeton University in the fall of 1966 and has thus benefited from a testing in the classroom.

The illustrations were prepared by John Kyrk. The typescript and much of the bibliography and index were prepared by Kathleen Horton with the assistance of Muriel Randall. Our personal research projects have been generously supported from the beginning by grants from the National Science Foundation and our respective home institutions.

R. H. M.
Department of Biology
Princeton University

E. O. W.
Department of Biology
Harvard University

December 15, 1966

Symbols Used

A	Area of an island
B	The age of an organism at which first offspring are produced (see the fecundity model of Chapter 4).
b_x, b_τ	The number of offspring born to an individual at age x or at age τ.
D (and d).	The distance between two islands.
E	The extinction rate, in species per unit time. The particular value of E found when the biota is in equilibrium is labelled X (q.v.).
F_i	The density of individuals in prey populations above which the predator species specializing on them can increase (Chapter 5).
G	In the turnover model for single islands (Chapter 3), G is used to designate $d\mu/dS - d\lambda/dS$, in other words, the difference in the slopes of the extinction curve and the immigration curve.
I	The immigration rate, in species per unit time.
J	The total number of individuals in a taxon at a given time.
K	The "carrying capacity of the environment," i.e., the number of individuals in a population of a given species at the population equilibrium. A population with more than K individuals will decrease.
Λ	The mean overseas dispersal distance of propagules of a given species.
λ	Used without subscript, this letter symbolizes the per capita birth rate, measured in individuals per individual per unit time.
λ_S	The rate of immigration of new species when S species are present.

λ_x The rate of birth of organisms when x organisms are present, measured in individuals per unit time.

l_x, l_τ The probability of an organism's surviving to age x or to age τ.

μ Used without subscript, this letter symbolizes the per capita death rate, measured in individuals dying per individual per unit time.

μ_S The rate of extinction of species when S species are present.

μ_x The rate of death of organisms when x organisms are present, measured in individuals per unit time.

$n_0(t)$ The number of newborn individuals in a population between time t and time $t + 1$.

P (and p).   Used to designate the number of species in the "species pool," that is, the number capable of immigrating to the island whether they all survive or not.

P_i The density of individuals in a population of predator species i (Chapter 5).

P_S The probability of occurrence of a certain number (S) of species.

ρ Density of organisms, i.e., number of organisms per unit area.

R Number of propagules (as opposed to species) arriving on an island in a given unit of time.

r The "intrinsic rate of increase," the per capita rate of net increase in a given environment. (Mathematical explanation in Chapter 4.)

R_i The density of individuals in a population of prey species i (Chapter 5).

R_0 The replacement rate: the average number of female offspring left during her life by each female.

S Number of species.

$\hat{S}$ Number of species at equilibrium.

T In the fecundity model of Chapter 4, the age of greatest fecundity.

T_x — The average length of time before a population containing x individuals goes to extinction.

$T_{0.90}$ — The time required for a given taxon to reach 90% of the equilibrial number of species on an island.

$U(x)$ — The reproductive function of individual organisms: the probability of survival to age x times the number of offspring produced at age x.

v_x — The reproductive value: a measure of the expected number of offspring yet to be produced by an individual of a given age. In biogeographic terms, it may be defined as the expected number of individuals in a colony (at some remote future time) founded by a propagule of x-year-olds.

W — In the fecundity model of Chapter 4, the last age at which offspring are produced.

X — The extinction rate at species equilibrium.

x, y — Used generally to designate discrete numbers, with different meanings given in various equations.

z — The slope of the log-log plot of the area-species curve. The z value varies with the kind of area unit employed, e.g., square miles as opposed to hectares.

THE THEORY OF
Island
Biogeography

CHAPTER ONE

The Importance of Islands

"The Zoology of Archipelagoes," Charles Darwin wrote at an early moment in his career, "will be well worth examination."[1] And so it has proved. The study of insular biogeography has contributed a major part of evolutionary theory and much of its clearest documentation. An island is certainly an intrinsically appealing study object. It is simpler than a continent or an ocean, a visibly discrete object that can be labelled with a name and its resident populations identified thereby. In the science of biogeography, the island is the first unit that the mind can pick out and begin to comprehend. By studying clusters of islands, biologists view a simpler microcosm of the seemingly infinite complexity of continental and oceanic biogeography. Islands offer an additional advantage in being more numerous than continents and oceans. By their very multiplicity, and variation in shape, size, degree of isolation, and ecology, islands provide the necessary replications in natural "experiments" by which evolutionary hypotheses can be tested.

Insularity is moreover a universal feature of biogeography. Many of the principles graphically displayed in the Galápagos Islands and other remote archipelagos apply in lesser or greater degree to all natural habitats. Consider, for example, the insular nature of streams, caves, gallery forest, tide pools, taiga as it breaks up in tundra, and tundra

[1] As he left the Galápagos in 1835, Darwin was struck by the variation among the skins of mockingbirds he had just collected from the different islands. He then wrote in his notebook what is considered to be the first reference to an awakening interest in evolution, as well as the first glimpse of modern island biogeography: "When I see these Islands in sight of each other, & possessed of but a scanty stock of animals, tenanted by these birds, but slightly differing in structure & filling the same place in Nature, I must suspect they are only varieties . . . . If there is the slightest foundation for these remarks the Zoology of Archipelagoes will be well worth examination; for such facts would undermine the stability of species."

as it breaks up in taiga. The same principles apply, and will apply to an accelerating extent in the future, to formerly continuous natural habitats now being broken up by the encroachment of civilization, a process graphically illustrated by Curtis's maps of the changing woodland of Wisconsin seen in Figure 1.

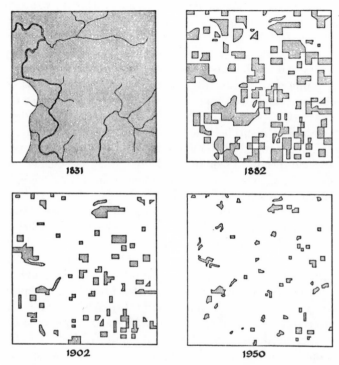

FIGURE 1. Reduction and fragmentation of the woodland in Cadiz Township, Wisconsin, 1831–1950. (After Curtis, 1956.)

Biogeography is a subject hitherto little touched by quantitative theory. The main reason is that the fundamental processes, namely dispersal, invasion, competition, adaptation, and extinction, are among the most difficult in biology to study and to understand. Stating postulates for even the simplest models is a risky business, because we are unsure of the complex biological phenomena underlying them. Another reason is that most research has been taxonomic in origin, and dominated by the historical viewpoint. The conventional issues relate to special places and special groups of plants and animals. The major questions are *ad hoc* and historically oriented; for example:

What was the ultimate origin of the Antillean vertebrate fauna?

Did Central America develop a discrete insular fauna during the Tertiary?

How can we account for the phylogenetic similarities of the biotas of southern South America and New Zealand?

Why is Hawaii rich in species of *Nesoprosopis* but lacking in other native bee genera?

Partly because such questions are concerned with a limited number of higher taxa, and partly because of the considerable intrinsic interest in these taxa in the first place, the historical solutions have tended to be satisfying in themselves and have not encouraged generalizations.

The purpose of this book is to examine the possibility of a theory of biogeography at the species level. We believe that such a development can take place by looking at species distributions and relating them to population ecological concepts, both known and still to be invented. Although such formulations will be crude at first and perhaps often fall short of the intended goals in particular cases, the effort deserves to be made, for the following reason. A theory attempts to identify the factors that determine a class of phenomena and to state the permissible relationships among the factors as a set of verifiable propositions. A purpose is to simplify our education by substituting one theory for many facts. A good theory points to possible factors and relationships in the real world that would otherwise remain hidden and thus stimulates new forms of empirical research. Even a first, crude theory can have these virtues. If it can also account for, say, 85% of the variation in some phenomenon of interest, it will have served its purpose well. We need to ask next whether biogeography has a solid enough empirical basis at the present time to make such an

5

attempt. Certainly the amount of information on distribution is vast; it has been created by two hundred years of accumulated descriptive taxonomy. But data of use to a population theory of distribution are quite scarce. A main goal of this book is to identify those kinds of data needed for a further development of a population theory and, ultimately, the full explanation of distribution itself.

TABLE 1. Interrelations of chapters

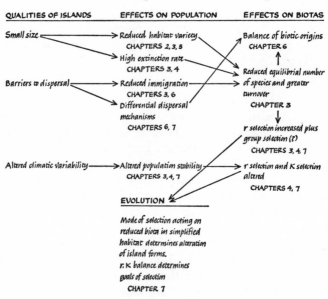

QUALITIES OF ISLANDS	EFFECTS ON POPULATION	EFFECTS ON BIOTAS
Small size	Reduced habitat variety CHAPTERS 2, 3, 5	Balance of biotic origins CHAPTER 6
	High extinction rate CHAPTERS 3, 4	Reduced equilibrial number
Barriers to dispersal	Reduced immigration CHAPTERS 3, 6	of species and greater turnover
	Differential dispersal mechanisms CHAPTERS 6, 7	CHAPTER 3
		r selection increased plus group selection (?) CHAPTERS 3, 4, 7
Altered climatic variability	Altered population stability CHAPTERS 3, 4, 7	r selection and K selection altered CHAPTERS 4, 7
	EVOLUTION	
	Mode of selection acting on reduced biota in simplified habitat determines alteration of island forms. r, K balance determines goals of selection CHAPTER 7	

In the chapters to follow we begin with a consideration of one of the more strikingly orderly relations encountered in biogeography: the area-diversity curve. Starting with the known facts concerning this relation, a rather extensive theory of the equilibrium of species is developed. The theory leads first to a consideration of the influence of life-table parameters of individual organisms on the immigration and extinction rates of populations and then to generalizations about the evolutionary strategies species must adopt in order to be good colonizers. The role of stepping stones in dispersal and exchange of species is next analyzed. Finally, some consequences of island biogeography on evolutionary theory are described. In order to assist the reader in visualizing the connections between the several major topics, a diagram is given in Table 1 which shows the sequences of causation as we conceive them and their location as topics in the next six chapters of the book.

CHAPTER TWO

Area and Number of Species

Theories, like islands, are often reached by stepping stones. The "species-area" curves are such stepping stones. Our ultimate theory of species diversity may not mention area, because area seldom exerts a direct effect on a species' presence. More often area allows a large enough sample of habitats, which in turn control species occurrence. However, in the absence of good information on diversity of habitats, we first turn to island areas.

There exists within a given region of relatively uniform climate an orderly relation between the size of a sample area and the number of species found in that area. Darlington (1957; see our Figure 2) expressed it as follows for the particular case of the herpetofauna of the West Indies: "division of area by ten [in going from one island to the next] divides the fauna by two." A more general first approximation for number of species in island faunas as a whole is given by the equation $S = CA^z$, where S is the number of species of a

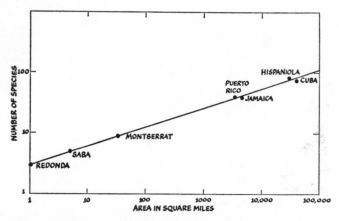

FIGURE 2. The area-species curve of the West Indian herpetofauna (amphibians plus reptiles).

8

given taxon found on the island and A is the area of the island. C is a parameter that depends on the taxon and biogeographic region, and in particular most strongly on the population density determined by those two parameters. z is also a parameter but one that changes very little among taxa or within a given taxon in different parts of the world. Of course if area is correlated with elevation and proximity to the mainland, z may be increased. Actual values of z are given in Table 2. Most cluster in the range 0.20–0.35;

TABLE 2. The z values of various taxa of land plants and animals on archipelagos. (From the equation $S = CA^z$, where S is the number of species and A is area in square miles.)

Fauna or flora	Island group	z	Authority
Carabid beetles	West Indies	0.34	Darlington (1943)
Ponerine ants	Melanesia	0.30	Wilson (original)
Amphibians and reptiles	West Indies	0.301	Preston (1962)
Breeding land and fresh-water birds	West Indies	0.237	Hamilton et al. (1964)
Breeding land and fresh-water birds	East Indies	0.280	Hamilton et al. (1964)
Breeding land and fresh-water birds	East-Central Pacific	0.303	Hamilton et al. (1964)
Breeding land and fresh-water birds	Islands of Gulf of Guinea	0.489	Hamilton and Armstrong (1965)
Land vertebrates	Islands of Lake Michigan	0.239	Preston (1962)
Land plants	Galápagos Islands	0.325	Preston (1962)

for example, the z value obtained arithmetically from Darlington's rule of thumb is 0.301, which is very close to the measured value for the West Indian herpetofauna. A notable exception, the value of 0.489 for the land and fresh-water birds of the islands of the Gulf of Guinea, may in fact not be representative, since it was based on only four islands.

When we shift from islands to non-isolated sample areas *within* islands, or within continents, a similar relation between area and species number exists, except that now z is smaller, falling usually between 0.12 and 0.17. An example from the ants is given in Figure 3. This particular relationship is of importance to descriptive ecology and has been dealt with in consummate detail by Preston (1962) and Williams (1964). The reason for the lower z values will be discussed shortly.

Preston has made an important contribution by demonstrating that the z values commonly encountered in island

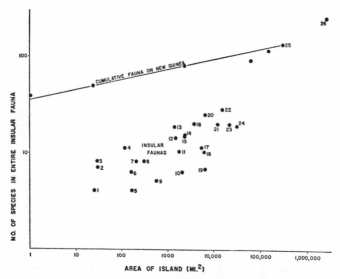

FIGURE 3. The relationship of area to number of ponerine (including cerapachyine) ant species in the faunas of various Moluccan and Melanesian islands. The line and the cluster of points illustrate the principle that the increase in number of species with area is more rapid in the case of isolated islands or archipelagos than in expanding sample areas on a single land mass. Ternate, 1; Malapaina, 2; Ugi, 3; Florida, 4; Kandavu, 5; Taviuni, 6; Ndeni, 7; Amboina, 8; Rennell, 9; Vanua Levu, 10; Espiritu Santo, 11; San Cristoval, 12; Santa Isabel, 13; Malaita, 14; Waigeo, 15; Viti Levu, 16; New Hebrides (entire), 17; Ceram, 18; Halmahera, 19; Fiji (entire), 20; New Britain, 21; Solomons (entire), 22; Bismarck Archipelago (entire), 23; Moluccas, 24; New Guinea, 25; central tropical Asia, 26. The cumulative New Guinea localities given in the curve ascend as follows: lower Busu River; triangle formed by the lower Busu River, Didiman Creek, and Bubia; all of the Huon Peninsula; northeast New Guinea; northeast New Guinea plus Papua; all of New Guinea. (From Wilson, 1961.)

10

biotas can be approximated *a priori* by assuming the relative abundances of the species to be distributed in a lognormal fashion, a possibility also mentioned by Goodall (1952). Since the local faunas (or floras) do typically exhibit abundance curves at least approaching the lognormal distributions, Preston's derivation links two independent generalizations of ecology, namely the typical forms of the species-abundance curve and the area-species curve. The formulation is therefore worth examining in detail.

The lognormal distribution is a simple concept. When we plot (on the abscissa) the absolute number of individuals per species against the number (on the ordinate) of species falling in each abundance class, the result is a frequency curve skewed strongly away from the lower values on the left. That is to say, there are many more moderately rare species than moderately common ones. Preston calls the distribution the Species Curve; it has also been referred to in the literature as the species-abundance curve. When the abscissa is changed to a logarithmic scale, the curve becomes symmetrical or nearly so. This approximation can be expressed analytically in various ways. Preston used a scale of "octaves" on the abscissa, an octave being an interval in which the number of individuals per species is doubled. Thus the first few octaves are 1–2, 2–4, 4–8, 8–16, . . . , etc., with species having boundary abundances of precisely 2, or 4, or 8, or 16, *et seq.* being split into the two adjacent octaves. Preston found that the number of species containing a certain number of individuals could be roughly predicted from the equation

$$y = y_0 e^{-(aR)^2}$$

where y is the number of species to be found in the Rth octave to the right or left of the mode, y_0 is the number of species in the modal octave, and a is a fitted constant. The constant a is related to the logarithmic standard deviation σ by the equation $a^2 = \frac{1}{2\sigma^2}$. The Species Curve is an approximation of real curves obtained by censusing local mainland bird faunas.

A second curve, called the Individuals Curve, measures the number of individuals (rather than species) found in the various species-abundance classes. Hence the height of the Individuals Curve above a given octave is y times the mean abundance of individuals in that octave. If the Species Curve reaches its mode at n_0 individuals per species, there are $n_0 2^R$ individuals per species in the Rth octave right of the mode and there are y such species. Multiplying, the number of individuals R octaves right of the mode is given by

$$ Y = n_0 y_0 2^R e^{-a^2 R^2} = n_0 y_0 \exp\left(\frac{\ln 2}{2a}\right)^2 \exp\left[-a^2\left(R - \frac{\ln 2}{2a^2}\right)^2\right]. $$

This is itself a normal distribution, displaced $(\ln 2)/2a^2$ octaves right of the Species Curve. The Individuals Curve has a tendency to terminate near its own mode, and for purposes of simplification Preston assumed that it terminates precisely on the mode. This is Preston's "canonical" hypothesis. Put another way, the number of species is assumed to become vanishingly small $(\ln 2)/2a^2$ octaves to the right of the mode. This point is $\sigma^2 \ln 2$ octaves from the Species Curve mode. The Species Curve must drop to about $\frac{1}{2}$ at this point, in order that the probability of having any species be small. From tables of the normal distribution we can thus calculate σ theoretically (see Table 3). The two types of curves are illustrated in an artificial example in Figure 4.

Taking the general forms of the two types of curve as postulates, it becomes possible to relate the number of species in the ensemble not only to the theoretical a, σ, and empirical y_0, but also to $\pm x$, the range in octaves on either side of the mode which on the average will embrace all the species, i.e., the value such that at $x\sigma$ octaves from the mode, $y = \frac{1}{2}$; and then to J/m, the ratio of the total number of individuals (J) in the ensemble to the number of individuals (m) in the rarest species. As J increases, by an increase either in area or in population density, x increases; hence S increases as more species can be fitted in under the rising lognormal curve. Because of the presumed close inter-

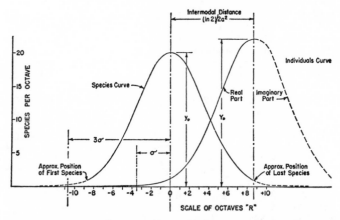

FIGURE 4. The "canonical" distribution for 178 imaginary species of an insular fauna according to the formulation of Preston. For this number the coefficient $a = 0.200$ and the standard deviation σ is 3.53 octaves for both the Species Curve and Individuals Curve. The modal height of the Species Curve is $y_0 = 20$ species; the intermodal distance is 8.68 octaves. The real part of each curve is drawn solid. The first (rarest) and last (commonest) species should fall at about this distance (8.68 octaves or 2.45 σ) from the mode of the Species Curve. (From Preston, 1962.)

dependence of these variables, Preston refers to the entire structure as "canonical." The predicted relationships are given in Table 3. From these estimates, the following interesting relation can be computed by least-squares fitting:

$$\log J/m = 3.821 \log S - 1.21,$$

or by rearrangement

$$\log S = 0.263 \log J/m + 0.317.$$

Now if the set of islands occurs in an area that is more or less uniform in climate and topography, we can assume that J, the number of individual organisms in the taxon on a given island, increases linearly with the area of the island in the form

$$J = \rho A,$$

where ρ is the density of the individual organisms. Substitut-

13

TABLE 3. The parameters of Preston's "canonical distribution," which can be used to predict species numbers and abundances in an insular biota. Once J/m is given, the other parameters are fixed—including the number of species, S.

S	x	σ	a	y_0	J/m
100	2.576	3.72	0.190	10.7	2.69×10^6
200	2.807	4.05	0.175	19.7	3.75×10^7
400	3.042	4.36	0.162	36.6	4.82×10^8
800	3.227	4.66	0.152	68.5	6.23×10^9
1,000	3.291	4.75	0.149	84.0	1.47×10^{10}
2,000	3.481	5.02	0.141	158.9	2.16×10^{11}
4,000	3.662	5.28	0.134	302	2.75×10^{12}
8,000	3.836	5.54	0.128	576	4.02×10^{13}
10,000	3.891	5.61	0.126	711	0.93×10^{14}
100,000	4.418	6.38	0.111	6,130	6.61×10^{17}
1,000,000	4.892	7.06	0.100	56,500	5.21×10^{21}

S, number of species.

x, range in octaves on either side of mode of Species Curve which embraces all species in most cases.

σ, standard deviation of Species Curve in logarithmic units.

a, fitted constant (see text).

y_0, number of species in modal octave.

J, total number of individual organisms.

m, number of individual organisms in rarest species, i.e., the threshold population size for survival.

ing this expression in the previous equations and then treating m as an undetermined constant we get

$$S = 1.83(\rho/m)A^{0.263},$$

or

$$S = CA^{0.263}.$$

The predicted z value[1] of 0.263 falls toward the lower end of the empirically determined z values listed in Table 2.

The application of Preston's theory to islands requires that the canonical hypothesis and, more generally, the lognormal distribution of abundance, apply equally to mainland and islands. We give here two real examples of relative

[1] Not 0.262 as given by Preston. Actually, as pointed out by Preston, the relation of J/m to S is not exactly linear; and over the range 100 to 1,000 for S, which embraces the numbers used in some empirical studies, the z value is closer to 0.27.

14

abundance curves fitted to truncated lognormal distributions as studied by Dr. Ruth Patrick, who has kindly let us use her unpublished data. In Figure 5 a comparison is made of the relative abundances of diatoms in chemically similar streams on the island of Dominica and on the American mainland. The actual curves hardly seem close enough in

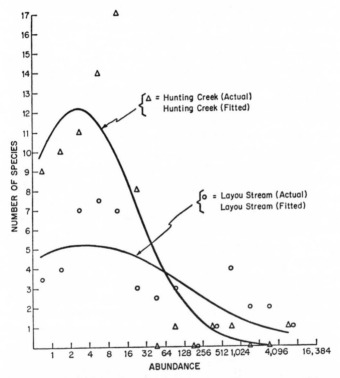

FIGURE 5. Truncated lognormal curves for diatom communities calculated according to the method of Preston. The Patrick paper from which these data were taken is in manuscript form, soon to be published. These curves support the assumption that species-abundance curves are close to the lognormal form in both continental and island communities. △—△ Hunting Creek, Maryland; ○—○ Layou Stream near Clark Hall, Island of Dominica. Notice how on Dominica, where fewer species are present, the standard deviation, σ, of the fitted, normal curve is much greater (a is smaller). This aspect does not conform to Preston's hypothesis.

15

form to each other and to the ideal lognormal form to support the basic assumption.

Deviations from the theoretical value z are to be expected, and should be subject to biological explanation. Consider first the case already mentioned of non-isolated sample areas on continents or large islands, which tend to exhibit z values between 0.12 and 0.17, well below the range of the parameter on islands. A small sample area carries more species than an island of the same size and similar environment probably for a quite simple reason. It contains very small numbers of individuals—perhaps only single individuals in extreme cases—belonging to species that are not well adapted to the sample area but are nonetheless represented because they persist in other places close at hand. In other words, there is a much higher immigration rate of transient species than is the case in the more isolated islands. Expressed in terms of Preston's formulation, m is decreased, and J/m and S are correspondingly increased. This advantage to non-isolated, continental sample areas diminishes, however, toward the upper end of the area scale. As the area is enlarged, it envelops an ever more complete sample of the habitats on the continent as a whole. As a result the number of alien habitats in surrounding areas decreases, and the proportion of transient species correspondingly decreases. Consequently the species-area curve ascends less rapidly, that is, z is smaller. In comparing archipelagos, we would expect that as a rule z will increase with increasing degree of isolation, both of the islands from each other and of the archipelago as a whole from the nearest neighboring archipelago or continent. Hamilton and Armstrong (1965) have pointed out that this is true for land birds of the West Indies, East Indies, and East-and-Central Pacific, which are increasingly isolated in that order. It is not true for the islands of the Gulf of Guinea; but in this case, as we have seen, the sample size may not be large enough to be meaningful. Later we will show how the same result can be predicted in a different way from equilibrium models.

Now let us consider how z values can be pushed higher than the predicted value of 0.26 or 0.27. As islands become large, their topography becomes more complex, especially if they are mountainous. The result is a growing heterogeneity of habitats, each of which can support ensembles of species that are ecologically semi-independent of each other. As a result, not only is the total number of individuals increasing and advancing the number of species in the expected canonical manner, but in addition the island as a whole is breaking up into multiple "semi-islands," such as lowland rain forests, montane rain forests, rain-shadow grasslands, and other principal habitats, whose species can evolve at least to some degree as independent assemblages. The result is an enhancement of species accumulation that will account for some, and perhaps most, of the observed increment of insular z values above the predicted number.

Finally the coefficient C of the area-diversity equation deserves an additional brief consideration. This parameter, which depends so much on the population density as well as on the innate species diversity of the given taxon, can be expected to vary greatly among taxa and also, within the same taxon, in different parts of the world. In the latter case, C clearly should be less in those regions where the quality of the environment is poorer and the total number of organisms in the taxon is smaller. C should also decrease with increased isolation. Beyond making these obvious broad generalizations, it does not seem possible at this time to develop any general theory that would permit precise predictions of C values.

SUMMARY

The number of species on a given island is usually approximately related to the area of the island by the equation $S = CA^z$, where S is the number of species, A is the area, C is a constant that varies widely among taxa and according to the unit of area measurement, and z is a constant which falls in most cases between 0.20 and 0.35. The empirically determined values of z are consistent with the independently derived generalization that the frequency curve of species containing various numbers of individuals (the latter divided into abundance classes on the abscissa) are distributed lognormally. As first shown by Preston, the z value resulting from a lognormal distribution should be 0.27. When, however, the species counts are made from sample plots of increasing area on the *same* island or continent, the z values are smaller, usually falling between 0.12 and 0.17. This deviation can be at least partly explained as the result of flooding of small sample plots by transient species that maintain themselves in ecologically different areas nearby— an exchange that is drastically reduced among islands due to the barriers that separate them. The opposite deviation, an increase in z above 0.27, can be explained as the outcome of the breaking up of biotas of large islands into semi-isolated communities due to the increase in topographic barriers and environmental variation on such islands. The one example of relative abundance curve from islands conforms in some, but not all, aspects to Preston's hypothesis.

CHAPTER THREE

Further Explanations
of the Area-Diversity Pattern

The Concept of the Species Equilibrium

The strikingly orderly relation between island area and species diversity has elicited several attempts to identify and to measure the contributing factors. Koopman (1958) selected a fauna—the bats of the southern Caribbean islands —in which distance could be minimized as an influence, and used its geography to illustrate the important roles played by both area and ecology in determining species diversity. His data, summarized in Figure 6, show how much major differences in ecology among islands can distort the area-diversity curve. Because such differences do occur over short distances in many parts of the world, area alone cannot be assumed in any particular case to be a precise predictor of species diversity.

Hamilton *et al.* (1964, 1967) and Watson (1964) have employed the more precise method of multiple regression analysis to the same end in studies of island bird faunas. In particular they have measured the effects of degree of isolation, island area, island elevation, and in one case (Watson) a measure of island complexity. The technique is most useful when considering the properties of single archipelagos. For example, in their analysis of bird diversity of individual tropical archipelagos, Hamilton, Barth, and Rubinoff (1964) and Hamilton and Rubinoff (1967) found that in most cases area accounts for 80–90% of the variation and elevation for another 2–15%; for reasons that are not yet clear, the Galápagos form an exception in that degree of isolation is the dominant factor and area only a minor one.

While multiple regression analysis, or some equivalent quantitative analysis, is indispensible for sorting out com-

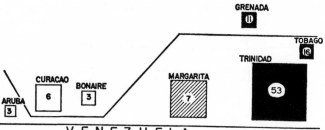

FIGURE 6. In this diagrammatic representation of the islands off the coast of South America, the numbers of species of bats are shown to be determined primarily by island area and ecology, rather than by degree of geographic isolation. The size of each square is proportional to the size of the indicated island. The shading of each square indicates the vegetation type of the island (black, rain forest; cross-hatched, traces of rain forest; white, xerophytic vegetation only). The numbers given are the numbers of species recorded from each island. The straight line near the bottom represents the Venezuelan coastline, the zigzag line above it the 100-fathom line. Notice that Grenada, which is both small and isolated, still has four more species than the large, near island of Margarita, and many times more species than the small, near islands of Aruba and Bonaire. The critical factor in this case is ecology. (From Koopman, 1958.)

ponents that are not relevant, it is not a very fruitful way of generating new hypotheses. Part of the difficulty is that neither area nor elevation exerts a direct effect on numbers of species; rather, both are related to other factors, such as habitat diversity, which in turn control species diversity. Watson, by separating out a measure of island habitat diversity in the Aegean islands, found that it accounted for the greatest part of variation in number of resident bird species, leaving little residual distance or area effect. Yet even this more definitive correlation does not in itself suggest new explanatory hypotheses about the control of species diversity.

Preston (1962) and MacArthur and Wilson (1963) independently suggested that there might be a balance of immigration by extinction so that the diversity of at least some biotas could be understood as an equilibrium. A perfect balance between immigration and extinction might never be reached, since it would be approached exponentially; but to the extent that the assumption of a balance has enabled us to make certain valid new predictions, the equilibrium concept is useful as a step beyond the more purely descriptive techniques of multiple regression.

Figure 7 shows the basic idea of the equilibrium condition.

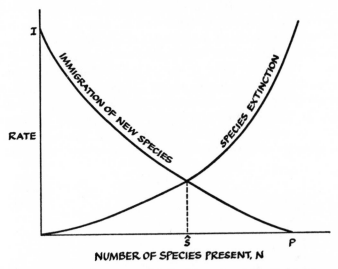

FIGURE 7. Equilibrium model of a biota of a single island. The equilibrial species number is reached at the intersection point between the curve of rate of immigration of new species, not already on the island, and the curve of extinction of species from the island. (After MacArthur and Wilson, 1963.)

Both the immigration and extinction rates vary with the number of species present. The immigration rate (in new species per unit time) is a falling curve because as more species become established, fewer immigrants will belong to new species. We expect the curve to be concave, i.e., steeper at the left, because on the average the more rapidly dispersing species would become established first, causing a rapid initial drop in the overall immigration rate, while the later arrival of slow colonizers would drop the overall rate to an ever diminishing degree. (A more complete analysis is presented later in this chapter.) The extinction curve must on the other hand rise, since the more species there are present the more there are to become extinct, and the more likely any given one will become extinct due to the smaller average population size acting through both ecological and genetical accident. We have moreover assigned the extinction curve an approximately exponential shape, reasoning that the combination of diminishing population size and increasing probability of interference among species will have an accelerating detrimental effect. Yet these refinements in shape of the two curves are not essential to the basic theory. So long as the curves are monotonic, and regardless of their precise shape, several new inferences of general significance concerning equilibrial biotas can be drawn. For example, as shown in Figure 8, where immigration is reduced or extinction increased, the island will equilibrate at fewer species. Just this result should obtain,

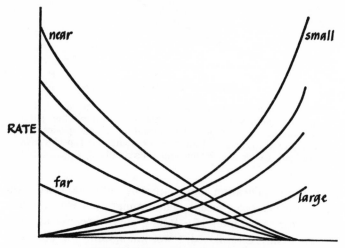

NUMBER OF SPECIES PRESENT, N

FIGURE 8. Equilibrium models of biotas of several islands of varying distances from the principal source area and of varying size. An increase in distance (near to far) lowers the immigration curve, while an increase in island area (small to large) lowers the extinction curve. (After MacArthur and Wilson, 1963.)

therefore, where the degree of isolation from the source regions that supply the species is increased, reducing immigration, or where the area of the island is decreased, resulting in a higher extinction rate.

The model accounts for the "impoverishment" of distant insular biotas. But the same effect can be predicted from a non-equilibrial hypothesis which states that the distant islands have not yet filled up because of the lower immigration rate supplying them. A stronger test for the equilibrial condition is fulfillment of the second condition deduced from the model; namely that the logarithms of the species numbers increase with area more rapidly on distant islands than on near islands. As MacArthur and Wilson (1963) showed, the impoverished bird faunas of central and eastern Polynesia seem to meet this condition. The predicted relationships are displayed by the data in Figures 9

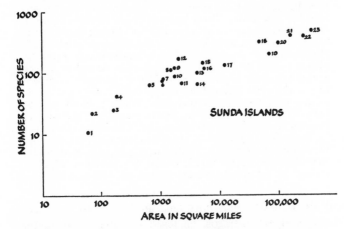

FIGURE 9. The numbers of land and fresh-water bird species on various islands and archipelagos of the Sunda group, together with the Philippines and New Guinea. The islands and archipelagos are grouped close to one another and to the Asian continent and Greater Sunda group, where most of the species live; and the distance effect is not apparent. Christmas, 1; Bawean, 2; Engano, 3; Savu, 4; Simalur, 5; Alors, 6; Wetar, 7; Nias, 8; Lombok, 9; Billiton, 10; Mentawei, 11; Bali, 12; Sumba, 13; Bangka, 14; Flores, 15; Sumbawa, 16; Timor, 17; Java, 18; Celebes, 19; Philippines, 20; Sumatra, 21; Borneo, 22; New Guinea, 23. (Modified from MacArthur and Wilson, 1963.)

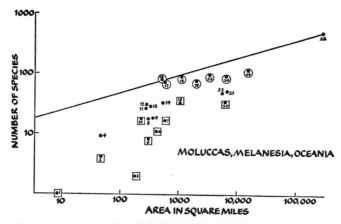

FIGURE 10. The numbers of land and fresh-water bird species on various islands and archipelagos of the Moluccas, Melanesia, Micronesia, and Polynesia. Here the archipelagos are widely scattered, and the distance effect is apparent in the greater variance. Hawaii is included even though its fauna is derived mostly from the New World. "Near" islands (less than 500 miles from New Guinea) are enclosed in circles, "far" islands (greater than 2,000 miles) are enclosed in squares, and islands at intermediate distances are left unenclosed. A line is drawn through two of the islands nearest the source regions and with the highest species densities in order to give a clearer idea of the degree of departure of the other islands from the potential densities of archipelagos with high immigration rates. Wake, 1; Henderson, 2; Line, 3; Kusaie, 4; Tuamotu, 5; Marquesas, 6; Solomons, 7; Ponape, 8; Marianas, 9; Tonga, 10; Carolines, 11; Palau, 12; Santa Cruz, 13; Rennell, 14; Samoa, 15; Kei, 16; Louisiade, 17; D'Entrecasteaux, 18; Tanimbar, 19; Hawaii, 20; Fiji, 21; New Hebrides, 22; Buru, 23; Ceram, 24; Solomons, 25; New Guinea, 26. (Modified from MacArthur and Wilson, 1963.)

and 10. In particular, note that in Figure 10 the distant islands and archipelagos, enclosed in squares, do increase in species numbers more rapidly over at least much of the range in area. Even so, this evidence still cannot be accepted as proof of the equilibrium model. The number of distant faunas available for comparison are few in number, and at least one (Line Islands, number 3 in Figure 10) is impoverished at least in part because of an unfavorably dry environment. Thomas Schoener (personal communication) has further pointed out to us that the distant-fauna curve must

24

inevitably have a decreased slope in the upper part of the area range. As indicated by the position of Hawaii (number 20), it could bow downward so far as to have an even lower slope than the near-fauna curve in this part of the area range. Schoener has found by closer analysis of published faunal surveys that when separate area-diversity curves are drawn for sets of islands *within* archipelagos, and not among archipelagos as done in Figure 10, the slopes of distant archipelagos are actually lower than those for near archipelagos. Schoener's result nevertheless need not conflict with the equilibrium model: the lower slopes can be accounted for as an outcome of clumping of islands, a perturbation which will be analyzed later (Figure 15). But clearly more area-diversity studies need to be made before the distance effect predicted by the equilibrium model can be tested. In particular, more points of the kind enclosed in squares in Figure 10 need to be secured. Perhaps these cannot be obtained in birds which show the distance effect only over hundreds of miles and hence on the relatively small number of very isolated oceanic islands. Biogeographers may have to turn to other groups that are both species-rich and more sensitive to geographic barriers in order fully to test and to extend the basic equilibrium model.

Thus, as it stands, the theory predicts that under a wide variety of conditions the area effect expressed in absolute numbers of species will be greater on more distant islands. This need not invariably be the case, as illustrated by the arrangement in Figure 11A, in which the immigration and extinction curves are deliberately constructed so as to make the effect greater on close islands. In other words $d\hat{S}/dA$ is often but not necessarily greater on distant islands. A more universal relation is that of $\log \hat{S}$ to A. In order to evaluate this relation we can take graphs in which both immigration and extinction curves are linear, as shown in Figure 11B. (The linear condition is not so stringent as it may seem, for any transformation of the ordinate is permissible, although not all immigration and extinction curves can be simultaneously straightened.) If the immigration and extinction

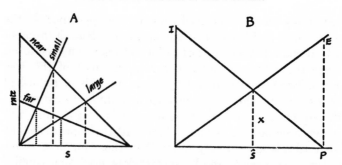

FIGURE 11. A: The equilibrium model with the immigration and extinction curves arranged so that the increment in species number is greater for near islands than for distant islands. This shows that the distance effect does not hold for all conditions when the scale of abscissa is in units of S (absolute number of species) rather than log S. B: Model with straightened immigration and extinction curves used to demonstrate the monotonic increase of the logarithm of the equilibrial number of species (log $\hat{S}$) with area. I, immigration rate at the beginning of colonization. X, extinction rate at equilibrium. $\hat{S}$, number of species at equilibrium. P, number of species in the species pool. E, extinction rate when P species are on the island.

curves are mirror images, then both may be straightened simultaneously by distorting the ordinate; otherwise our results will apply only where immigration and extinction curves are relatively straight. By similar triangles $X/\hat{S} = E/P$, so that $XP = E\hat{S}$, and also $X/(P - \hat{S}) = I/P$ whence $XP = I(P - \hat{S})$. Equating these values of XP,

$$E\hat{S} = I(P - \hat{S}) = IP - I\hat{S},$$

or

(3-1) $$\hat{S} = IP/(E + I).$$

We keep at one distance (I = const.) and consider variation in area, A, (or extinction, E) only. Now

$$\frac{d\hat{S}}{dE} = \frac{-IP}{(E + I)^2},$$

so that

$$\frac{d\hat{S}}{dA} = \frac{d\hat{S}}{dE} \cdot \frac{dE}{dA} = \frac{-IP}{(E + I)^2} \frac{dE}{dA}.$$

26

dE/dA is negative but independent of I. Changing I may increase or decrease this value, depending on E. It is easier to use log $\hat{S}$:

$$(3\text{-}2) \quad \frac{\mathrm{d} \log \hat{S}}{\mathrm{d}A} = \frac{1}{\hat{S}} \frac{\mathrm{d}\hat{S}}{\mathrm{d}A} = \left(\frac{E+I}{IP}\right)\left(\frac{-IP}{(E+I)^2}\right)\frac{\mathrm{d}E}{\mathrm{d}A}$$
$$= \frac{-\mathrm{d}E/\mathrm{d}A}{E+I}.$$

This *always* increases when the immigration rate, I, decreases. (Notice that the numerator is always positive.) In fact,

$$(3\text{-}2\mathrm{A}) \quad \frac{\mathrm{d}}{\mathrm{d}D}\left(\frac{\mathrm{d} \log \hat{S}}{\mathrm{d}A}\right) = \frac{(\mathrm{d}I/\mathrm{d}D)\cdot(\mathrm{d}E/\mathrm{d}A)}{(E+I)^2} > 0,$$

where D is distance. But if log $\hat{S}$ varies faster with area on distant islands we can rephrase the relation by saying log $\hat{S}$ varies faster with distance on small islands. This result has been derived specifically for linear immigration and extinction curves but doubtless is valid under more general conditions.

We next ask what happens when the species pool P varies instead of I (for instance, when an archipelago is colonized from a stepping-stone island). Then from the relation

$$\frac{\mathrm{d}\hat{S}}{\mathrm{d}A} = \frac{\mathrm{d}E}{\mathrm{d}A}\frac{\mathrm{d}\hat{S}}{\mathrm{d}E} = -\frac{\mathrm{d}E}{\mathrm{d}A}\frac{IP}{(E+I)^2},$$

we see that the rate of increase of $\hat{S}$ with area is greater for large P and less on archipelagos colonized via stepping stones. Curiously enough, this effect may not show if we plot log $\hat{S}$, for, returning to Equation (3-2) we find d log $\hat{S}$/dA is independent of P. For this effect we assume it is possible to alter P without changing I. Such a relation is plausible but needs empirical confirmation.

Now let us examine more closely the effects of changing area, distance, and configuration of the islands. First, consider the case illustrated in Figure 12 of two archipelagos, one near and one far, each containing large and small islands. In the formulation of Equation (3-2), dE/dA and E are the same for both archipelagos. The important feature is that

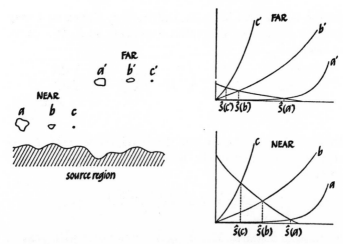

FIGURE 12. Diagram contrasting near and far archipelagos within which island area varies. The number of species increases with area more rapidly on the far archipelago. In this case the scale of the abscissa is in absolute number of species; the relationship becomes more nearly universal when the scale is made logarithmic.

when the archipelago is nearer the source region I is larger, the denominator is therefore larger, and the absolute value of the whole expression smaller. In other words, the closer archipelago will have an increased slope for its I curve and hence a reduced area effect. Thus qualitative differences in the species-area curves, such as those in near and distant Indo-Australian archipelagos illustrated in Figures 9 and 10, are among the results that can be obtained analytically as well as graphically.

Similarly, consider the situation in Figure 13 in which a set of islands of large, equal size but differing distances are contrasted with a second set of islands of small, equal size but differing distances. Now we must compute d log S/dD from Equation (3-1), and we find it to be larger when E is larger, i.e., A is smaller. Consequently the large-island sequence should show less distance effect on a logarithmic plot.

Notice that in Figures 12 and 13 the islands were placed

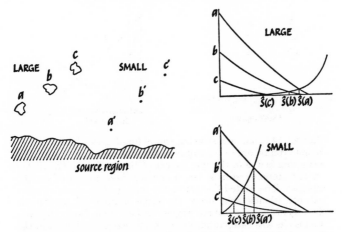

FIGURE 13. Diagram contrasting archipelagos with large and small islands, respectively, within which distance varies but not island area. The logarithm of the number of species decreases with distance more rapidly in the archipelago with small islands although the absolute number does not.

in such a way that colonists would not be likely to use the nearer ones as stepping stones. The equilibrium equations state the change in area effect with distance or the change in distance effect with area in terms of changes in the slopes of immigration and extinction curves. The critical result is that the immigration curve is less steep when it is less high at the left, *provided the intercept on the abscissa stays constant.* But stepping-stone colonization can be just the opposite: the outer island with a nearby stepping stone may have a numerically great rate of immigration of all species combined, but with only the few species from the stepping stones effectively participating. This immigration curve is steep; hence outer islands connected by stepping stones to the source region should show a reduced area effect when S is plotted against area (Figure 14).

A second perturbation can result from the influence of island clustering. The effect can be increased to the point where not only is the species pool available to the recipient islands increased so as to approach that of islands closer to the source region, but also the immigration curve is sub-

29

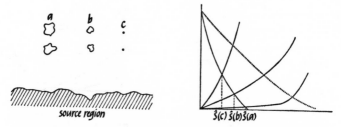

FIGURE 14. Diagram showing the reduced area effect in outer islands connected to the source regions by stepping stones. The reduction is due to the smaller size of the special pool available on the stepping stones, i.e., a smaller intercept value on the abscissa. The dotted line shows the immigration curve of comparable islands that draw the same number of propagules directly from the source region rather than from stepping stones. In this case, unlike other special cases considered in the text, the effect should hold universally true for the absolute number of species but not for the logarithm of the number of species.

stantially raised by the direct contribution of propagules from the stepping stones. As a result of the latter effect, I in Equation (3-2) is increased and d log $\hat{S}$/dA consequently decreased, i.e., there is a lowering of the slope of the area-species curve. The effect can be anticipated in archipelagos that are tightly clumped, as shown in Figure 15.

At least one truly anomalous feature of the area-species curve can be expected on extremely small islands. If the islands are so small as to be unstable, as is the case for example in the smallest keys of the Dry Tortugas (see Chapter 3), the turnover in biota can be rapid enough so that extinction rates are not area-dependent. Under this circumstance an increase in area at the lower end of the area scale would not result in an increase in species number. This relation, which is illustrated in Figure 16, seems the most economical hypothesis to offer for the peculiarities of Niering's data on the Kapingamarangi flora (Figure 17) and those of Wilson and Taylor (1967) on the Polynesian ants. Other, special explanations are not excluded and may even be consistent with the equilibrial hypothesis. For example, Wiens (1962) has suggested that the break in the

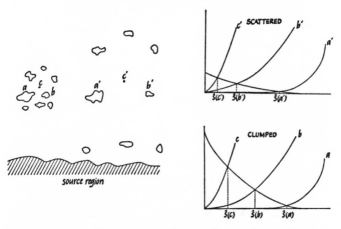

FIGURE 15. Predicted effect of increased clumping of islands. When clustered together, the islands raise each others' immigration rate, which in turn reduces the slope of the overall area-species curve.

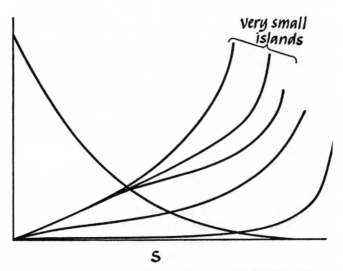

FIGURE 16. If species turnover on very small islands is great enough, extinction rates can be essentially independent of area. Consequently, in the lower end of the range in island area, the equilibrial species number will not increase with area, producing the condition shown in the Kapingamarangi flora curve of Figure 17.

31

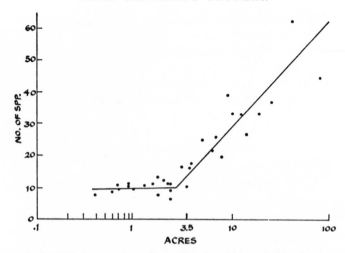

FIGURE 17. Area and number of species of higher plants in the islands of the Kapingamarangi Atoll, Micronesia. At the lower end of the range, species numbers do not increase with area. (Data from Niering, 1963.)

Kapingamarangi curve may be due to the occurrence of a fresh-water lens in islets of about 3.5 or more acres. If this proved to be true, the lens would increase the standing species number by reducing the extinction rates of species requiring fresh water.

Saturation and Turnover on a Single Island

The colonization of an island is a dynamic process, accompanied at all stages by turnover in species. During the buildup of species of a given taxon toward its equilibrial (saturation) number, the immigration and extinction rates probably vary as a function of the number of species present. It is reasonable to infer from this consideration alone that the variance of the number of species on different islands of a given size and degree of isolation will also vary with the number of species present; or, put the other way, the variance should be a function of the degree of saturation of the islands present. This idea can be developed as a formal model of the standard "birth and death process" type in probability theory.

Let $P_S(t)$ be the probability that, at time t, our island has S species, λ_S be the rate of immigration of new species onto the island when S are present, μ_S be the rate of extinction of species on the island when S are present; λ_S and μ_S then represent the intersecting curves of the graphical model (Fig. 7). This is a "birth and death process" only slightly different from the kind most familiar to mathematicians (cf. Feller, 1957, last chapter).

The reader who does not wish to study the details of the model at this time can pass on to the main conclusions given in Equation (3-14) and subsequent equations and text.

By the rules of probability

$$(3\text{-}3) \qquad P_S(t + h) = P_S(t)(1 - \lambda_S h - \mu_S h) \\ + P_{S-1}(t)\lambda_{S-1}h \\ + P_{S+1}(t)\mu_{S+1}h,$$

since to have S at time $t + h$ requires that at a short time preceding one of the following conditions held: (1) there were S and that no immigration or extinction took place, or (2) that there were $S - 1$ and one species immigrated, or (3) that there were $S + 1$ and one species became extinct. We take h to be small enough that probabilities of two or more extinctions and/or immigrations can be ignored. Bringing $P_S(t)$ to the left-hand side, dividing by h, and passing to the limit as $h \to 0$

$$(3\text{-}4) \quad \frac{dP_S(t)}{dt} = -(\lambda_S + \mu_S)P_S(t) + \lambda_{S-1}P_{S-1}(t) \\ + \mu_{S+1}P_{S+1}(t).$$

For this formula to be true in the case where $S = 0$, we must require that $\lambda_{-1} = 0$ and $\mu_0 = 0$. In principle we could solve Equation (3-4) for $P_S(t)$ because $\lambda_S = 0$ for S greater than the pool of species; for our purposes it is more useful to find the mean, $M(t)$, and the variance, var (t), of the number of species at time t. These can be estimated in nature by measuring the mean and variance in numbers of species on a series of islands of about the same distance and area and hence of the same λ and μ. To find the mean, $M(t)$, from Equation (3-4) we multiply both sides of (3-4) by S and then sum from $S = 0$ to $S = \infty$. Since

$$\sum_{S=0}^{\infty} SP_S(t) = M(t),$$

this gives us

$$\frac{dM(t)}{dt} = - \sum_{S=0}^{\infty} (\lambda_S + \mu_S)SP_S(t)$$

$$+ \sum_{S-1=0}^{\infty} \lambda_{S-1}[(S-1) + 1]P_{S-1}(t)$$

$$+ \sum_{S+1=0}^{\infty} \mu_{S+1}[(S+1) - 1]P_{S+1}(t).$$

(Here terms $\lambda_{-1} \cdot 0 \cdot P_{-1}(t) = 0$ and $\mu_0 \cdot (-1)P_0(t) = 0$ have been subtracted or added without altering values.) This reduces to

$$(3-5) \qquad \frac{dM(t)}{dt} = \sum_{S=0}^{\infty} \lambda_S P_S(t) - \sum_{S=0}^{\infty} \mu_S P_S(t)$$

$$= \overline{\lambda_S(t)} - \overline{\mu_S(t)}.$$

But, since λ_S and μ_S are, at least around their intersection point, approximately straight, the mean value of λ_S at time t is about equal to $\lambda_{M(t)}$ and similarly $\mu_S(t) \sim \mu_{M(t)}$. Hence, approximately

$$(3-6) \qquad \frac{dM(t)}{dt} = \lambda_{M(t)} - \mu_{M(t)},$$

or the expected number of species in the graphical model of Figure 7 moves toward $\hat{S}$ at a rate equal to the difference in height of the immigration and extinction curves. In fact, if $d\mu/dS - d\lambda/dS$, evaluated near $S = \hat{S}$ is abbreviated by G, then, approximately $dM(t)/dt = G[\hat{S} - M(t)]$ whose

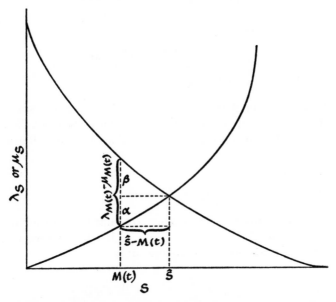

FIGURE 18. Diagram of immigration and extinction curves indicating the relationships used in deriving Equation (3-7). This figure is designed to provide a quicker intuitive grasp of the steps leading to later turnover equations in the text.

solution is $M(t) = \hat{S}(1 - e^{-Gt})$. This result can be obtained as follows. From the graph in Figure 18 and from Equation (3-6), note that

$$\frac{\mathrm{d}M(t)}{\mathrm{d}t} = \lambda_{M(t)} - \mu_{M(t)} = \alpha + \beta,$$

$$\frac{\mathrm{d}\lambda}{\mathrm{d}S} \doteq \frac{-\beta}{\hat{S} - M(t)},$$

$$\frac{\mathrm{d}\mu}{\mathrm{d}S} \doteq \frac{\alpha}{\hat{S} - M(t)}.$$

From this set of relationships and from the definition of G just given,

$$G = \frac{\mathrm{d}\mu}{\mathrm{d}S} - \frac{\mathrm{d}\lambda}{\mathrm{d}S}$$

$$= \frac{\alpha + \beta}{\hat{S} - M(t)}.$$

Then

$$dM(t)/dt = \alpha + \beta$$
$$= \frac{\alpha + \beta}{\hat{S} - M(t)} [\hat{S} - M(t)]$$
$$= G[\hat{S} - M(t)]$$

The solution for $M(t)$ is then verified by differentiating.

Finally, we can compute the time required to reach 90% (say) of the saturation value $\hat{S}$ so that $M(t)/\hat{S} = 0.9$ or $e^{-Gt} = 0.1$.

Therefore,

$$(3\text{-}7) \qquad t_{0.9} = 2.303/G$$

A similar formula for the variance is obtained by multiplying both sides of Equation (3-4) by $[S - M(t)]^2$ and summing from $S = 0$ to $S = \infty$. As before, since

$$\text{var}(t) = \sum_{S=0}^{\infty} [S - M(t)]^2 P_S(t),$$

this results in

$$(3\text{-}8) \quad \frac{d\,\text{var}(t)}{dt} = -\sum_{S=0}^{\infty} (\lambda_S + \mu_S)[S - M(t)]^2 P_S(t)$$

$$+ \sum_{S-1=0}^{\infty} \lambda_{S-1}\{[S - 1 - M(t)] + 1\}^2 P_{S-1}(t)$$

$$+ \sum_{S+1=0}^{\infty} \mu_{S+1}\{[S + 1 - M(t)] - 1\}^2 P_{S+1}(t)$$

$$= 2 \sum_{S=0}^{\infty} \lambda_S[S - M(t)]P_S(t)$$

$$- 2 \sum_{S=0}^{\infty} \mu_S[S - M(t)]P_S(t)$$

$$+ \sum_{S=0}^{\infty} \lambda_S P_S(t) + \sum_{S=0}^{\infty} \mu_S P_S(t).$$

Again we can simplify this by noting that the λ_S and μ_S curves are only slowly curving and hence in any local region are approximately straight. Hence, where derivatives are now evaluated near the point $S = M(t)$,

$$(3\text{-}9) \qquad \lambda_S = \lambda_{M(t)} + [S - M(t)]\frac{d\lambda}{dS}$$

$$(3\text{-}10) \qquad \mu_S = \mu_{M(t)} + [S - M(t)]\frac{d\mu}{dS}.$$

Substituting Equations (3-9) and (3-10) into (3-8) we get

$$(3\text{-}11) \quad \frac{d\,\mathrm{var}\,(t)}{dt} = 2(\lambda_{M(t)} - \mu_{M(t)}) \sum_{S=0}^{\infty} [S - M(t)]P_S(t)$$

$$+ 2\left(\frac{d\lambda}{dS} - \frac{d\mu}{dS}\right) \sum_{S=0}^{\infty} [S - M(t)]^2 P_S(t)$$

$$+ [\lambda_{M(t)} + \mu_{M(t)}] \sum_{S=0}^{\infty} P_S(t)$$

$$+ \left(\frac{d\lambda}{dS} + \frac{d\mu}{dS}\right) \sum_{S=0}^{\infty} [S - M(t)]P_S(t),$$

which, since

$$\sum_{S=0}^{\infty} P_S(t) = 1$$

and

$$\sum [S - M(t)]P_S(t) = M(t) - M(t) = 0$$

becomes

$$(3\text{-}12) \quad \frac{d\,\mathrm{var}\,(t)}{dt} = -2\left(\frac{d\mu}{dS} - \frac{d\lambda}{dS}\right) \mathrm{var}\,(t) + \lambda_{M(t)} + \mu_{M(t)}.$$

This is readily solved for var (t):

$$(3\text{-}12\mathrm{a}) \quad \mathrm{var}\,(t) = e^{-2[(d\mu/dS)-(d\lambda/dS)]t} \times \int_0^t (\lambda_{M(t)}$$
$$+ \mu_{M(t)})e^{2[(d\mu/dS)-(d\lambda/dS)]t}\,dt.$$

37

However, it is more instructive to compare mean and variance for the extreme situations of saturation and complete unsaturation, or equivalently of t = near infinity and t = near zero.

At equilibrium, [d var (t)]/dt = 0, so by Equation (3-12)

$$\text{(3-13)} \qquad \text{var }(t) = \frac{\lambda_{\hat{S}} + \mu_{\hat{S}}}{2\left(\dfrac{d\mu}{dS} - \dfrac{d\lambda}{dS}\right)}$$

At equilibrium $\lambda_{\hat{S}} = \mu_{\hat{S}} = X$ say and we have already symbolized the difference of the derivatives at $S = \hat{S}$ by G (cf. Equation (3-7)).

Hence, at equilibrium var $= X/G$. Now since μ_S has non-decreasing slope, $X/S \leq d\mu/dS|_{S=\hat{S}}$ or $X \leq \hat{S}\,d\mu/dS|_{S=\hat{S}}$. Therefore, variance

$$\leq \frac{\hat{S}\,d\mu/dS}{(d\mu/dS) - (d\lambda/dS)}$$

or, at equilibrium

$$\text{(3-14)} \qquad \frac{\text{variance}}{\text{mean}} \leq \frac{d\mu/dS}{(d\mu/dS) - (d\lambda/dS)}.$$

In particular, if the extinction and immigration curves have slopes about equal in absolute value, then the ratio (variance/mean) $\leq \frac{1}{2}$; but if the slopes are not equal the ratio would take some other value between 0 and 1. On the other hand, when t is near zero, Equation (3-12) shows that var $(t) \sim \lambda_0 t$. Similarly, when t is near zero, Equation (3-6) shows that $M(t) \sim \lambda_0 t$. Hence in a very unsaturated situation, approximately

$$\frac{\text{variance}}{\text{mean}} = 1.$$

Therefore, we would expect the variance/mean to fall from 1 to about $\frac{1}{2}$ as islands become saturated.

Finally, if the number of species dying out per year, X (at equilibrium), is known, we can estimate the time required to 90% saturation from Equations (3-7) and (3-13):

$$\frac{2.303}{t_{0.90}} = \frac{X}{\text{variance}}$$

(3-15) $\quad t_{0.90} = \dfrac{2.303 \text{ variance}}{X} = \dfrac{2.303}{2} \cdot \dfrac{\text{mean}}{X},$

the last depending on the slope of the immigration and extinction curves.

Qualitatively, then, the mean and variance of the number of species on a set of similar islands can be related to the steepness of the immigration and extinction curves and the ordinate and abscissa of their point of intersection. It would be possible to turn this procedure backwards to find X and G from the mean and variance, but we can actually go much further. We shall show that from the *distribution* of the number of species on a collection of similar islands we can come close to deducing the shape of the immigration and extinction curves. To do this, we write out Equation (3-4) explicitly for the equilibrium situation (where $dP_S(t)/dt = 0$ for all S), making use of the fact that $\lambda_S = 0$ when $S = p$, the species pool of immigrants, and that therefore $P_{p+1} = 0$:

$$0 = -\mu_p P_p + \lambda_{p-1} P_{p-1}$$
$$0 = -\lambda_{p-1} P_{p-1} - \mu_{p-1} P_{p-1} + \lambda_{p-2} P_{p-2} + \mu_p P_p$$
$$0 = -\lambda_{p-2} P_{p-2} - \mu_{p-2} P_{p-2} + \lambda_{p-3} P_{p-3} + \mu_{p-1} P_{p-1}$$

$$\vdots$$

$$0 = -\lambda_0 P_0 + \mu_1 P_1$$

From the first equation,

$$P_{p-1} = \frac{\mu_p}{\lambda_{p-1}} P_p.$$

Adding the top two of these equations,

$$P_{p-2} = \frac{\mu_{p-1}}{\lambda_{p-2}} P_{p-1} = \frac{\mu_{p-1}\mu_p}{\lambda_{p-2}\lambda_{p-1}} P_p.$$

Adding the top three equations,

$$P_{p-3} = \frac{\mu_{p-2}}{\lambda_{p-3}} P_{p-2} = \frac{\mu_{p-2}\mu_{p-1}\mu_p}{\lambda_{p-3}\lambda_{p-2}\lambda_{p-1}} P_p$$

and so on to

$$P_0 = \frac{\mu_1}{\lambda_0} P_1 = \frac{\mu_1 \cdots \mu_p}{\lambda_0 \cdots \lambda_{p-1}} P_p.$$

This gives us the exact equilibrium distribution of numbers

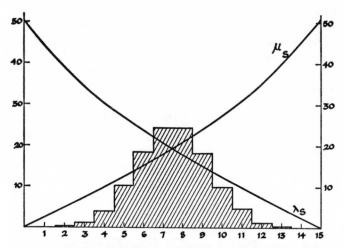

FIGURE 19. A particular case of a predicted distribution of numbers of species on a family of island biotas all with identical extinction and immigration curves and all having had time to reach equilibrium. The histogram represents the number of islands with each number of resident species in an equilibrium situation. The species pool from which the biotas were assembled contained 15 species. If the immigration and extinction curves were straighter, the variance of equilibrial species numbers would be even greater; yet this large variance is still consistent with the equilibrial condition.

of species on similar islands from the immigration and extinction curves (see Figure 19). The same equations can be solved for the *ratios* of extinction to immigration curves:

$$\frac{\mu_p}{\lambda_{p-1}} = \frac{P_{p-1}}{P_p}; \frac{\mu_{p-1}}{\lambda_{p-2}} = \frac{P_{p-2}}{P_{p-1}}; \cdots ; \frac{\mu_1}{\lambda_0} = \frac{P_0}{P_1}.$$

Hence when we know the values of P_0, P_1, . . . , P_p, then we can calculate the ratios of the values of the extinction and immigration curves.

Measurements of Colonization Rates and Turnover

True immigration and extinction rates should prove difficult not only to define but also to measure. Suppose by definition a propagule is a colonist as soon as it lands on an island. Then whether colonists live for long or not, we would, in order to compute the true immigration rate, find it necessary to monitor all incoming propagules, at various sea-

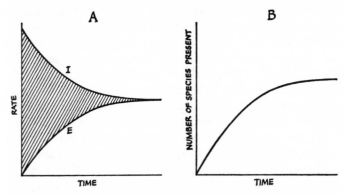

FIGURE 20. By integrating the difference through time of the immigration and extinction rates (diagram A), the colonization curve can be drawn (diagram B).

sons, in fluctuating weather, and at different times of the day. This to our knowledge has not been done in any single case, so that nothing approaching a reasonably accurate immigration curve can be drawn that relates the rate to the number of resident species. The same consideration applies to the measurement of extinction rate, which must involve the recording of failed propagules.

Figure 20 exemplifies how a *colonization curve* can be drawn by integrating the difference between the time curves of the immigration and extinction rates, if the latter two curves are known. The colonization curve is given by the equation

$$S_t = \int_0^t (I - E)\, \mathrm{d}t$$

We have already seen, following Equation (3-6), that the expectation, $M(t)$, of S_t has roughly the form $M(t) = \hat{S}(1 - e^{-Gt})$. Since the curve simply relates the rate of establishment of species to the number of species already present, it can also be drawn directly if the growth in numbers of species through time has been recorded. This is not difficult, providing the colonization rate is high enough so that a good part of the entire colonization process takes place over a period of at most a few years. Colonization curves can then be used to make indirect estimates of the immigration and extinction rates. Equation (3-15) gave the turnover rate (= extinction rate) as a function of the time required for the number of species to reach 90% of the equilibrial level. The latter number can be estimated from the colonization curve for the island together with the area-diversity curves of similar islands considered at equilibrium. If the colonizing taxon is known to have a high rate of success in new environments, the immigration rate at the beginning of the colonizing episode (when the island is empty of the taxon) should approach the colonization rate and can be estimated accordingly. Such estimates should be among the most readily obtained in field studies and can provide two points on both the immigration and extinction curves, one at the beginning and one near the end of the rise of the taxon to equilibrium.

Enough data are available to make the first approximations of both colonization and extinction curves. In Figures 21A and 21B are seen the results of some experiments by Maguire (1963a). For "islands" this experimenter used bottles of sterile, reconstituted fresh water placed at various distances from a fresh-water pond up to 400 yards and at various heights above the ground up to 4 feet. In the presence of a large but unmeasured immigration rate, the colonization curve drops off in agreement with the equilibrium model of our Figure 20. As Maguire points out, the micro-

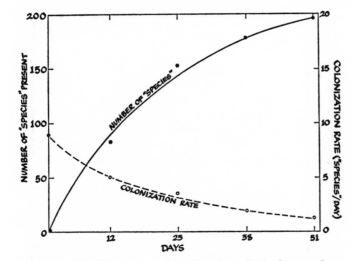

FIGURE 21A. Colonization of bottles of sterile fresh water by microorganisms visible at 45×. The source region was a fresh-water pond in Texas. The number of species given is not the true number. Only organisms that could be distinguished at 45× were included, and total species counts were simply added from several replications. Nevertheless the number is a fair measure of the species diversity. Notice that the form of the curve is consistent with the general form predicted by theory. (Based on Maguire, 1963a.)

organism species pool in Texas ponds is limited and his bottle faunulae must have approached it.

A second case can be drawn from the famous recoloniza-tion episode of the Krakatau Islands. The data on the growth of the bird faunulae of the islands, summarized by Dammerman (1948), provide an opportunity to test our stochastic model of the immigration and extinction processes on a single island. As is well known, the island of Krakatau proper exploded in August 1883, after a three-month period of repeated eruptions. Half of Krakatau disappeared en-tirely, and the remainder, together with the neighboring islands of Verlaten and Lang, was buried beneath a layer of glowing hot pumice and ash from 30 to 60 meters thick. Almost certainly the entire flora and fauna were destroyed. The repopulation proceeded rapidly thereafter. Collections

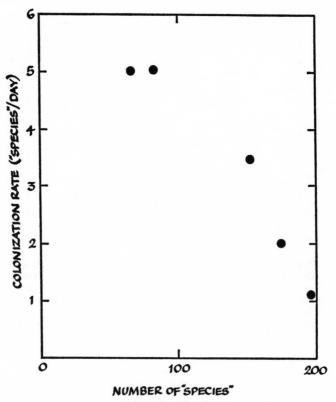

FIGURE 21B. The colonization rate of microorganisms as a function of numbers of species present, based on the colonization time curves of Figure 21A. (From Maguire, personal communication.)

and sight records of birds, made mostly in 1908, 1919–1921, and 1932–1934, show that the number of species of land and fresh-water birds on both Krakatau and Verlaten climbed rapidly between 1908 and 1919–1921 and did not alter significantly by 1932–1934 (see Table 4). Further, the number of non-migrant land and fresh-water species on both islands in 1919–1921 and 1932–1934, i.e., 27–29, fall very close to the extrapolated fauna-area curve of our Figure 9. (In fact, the area of each island after the 1883 eruptions was about 8 square miles, just outside the range shown in

TABLE 4. Number of species of land and fresh-water birds on Krakatau and Verlaten during three collection periods together with losses in the two intervals. (Based on data from Dammerman, 1948.)

| | 1908 | | | 1919–1921 | | | 1932–1934 | | | Number "lost" | |
	Non-migrant	Migrant	Total	Non-migrant	Migrant	Total	Non-migrant	Migrant	Total	1908 to 1919–1921	1919–1921 to 1932–1934
Krakatau	13	0	13	27	4	31	27	3	30	2	5
Verlaten	1	0	1	27	2	29	29	5	34	0	2

45

the figure.) A statement by Mayr (1965a), to the effect that comparable undisturbed faunas have many more species of birds, actually refers to the single, rather peculiar case of Durian, Riouw Archipelago, and is otherwise not well supported by the available data. Both lines of evidence suggest that the Krakatau faunas had approached equilibrium within only 25 to 36 years after the explosion. Although a precise colonization curve cannot be drawn on the basis of three isolated points, the rate was clearly dropping off in the later stages of the episode. The more rapid growth after 1908 presumably reflects the growth of a more diverse habitat, with trees present.

We can make use of the Krakatau data in two ways. First, supposing the fauna was virtually complete within 30 years, the equilibrium number of bird species is about 30, and from Equation (3-15) the extinction rate in species per year should be very roughly $X = (2.3 \times 30)/(2 \times 30) = 1.15$. To set a lower limit to the predicted turnover (= extinction) rate, let us suppose that Mayr (1965a) is closer in his estimate of the saturation level. Then the fauna was, after 30 years, only two-thirds saturated. From the equation for the expected number of species $M(t)$: $M(t) = \hat{S}(1 - e^{-Gt})$ we now get $(M(t))/\hat{S} = \frac{2}{3}$ when $e^{-Gt} = \frac{1}{3}$ or $t = 1.1/G$ instead of $2.3/G$. From Mayr's estimate, then, $X = 1.1 \times 42/2 \times 1/30 = 0.77$.

Of course these estimates assume that the slopes of immigration and extinction curves are equal. Any departure from equality would cause an appropriate change in the expected number, X, of extinctions per year. In any case, however, this number is astonishingly high; it is of the magnitude of 1% to 6% of the standing fauna. Yet the prediction seems to be supported by the collection data. On Krakatau proper, 5 non-migrant land and fresh-water species recorded in 1919–1921 were not recorded in 1932–1934, but 5 other species were recorded for the first time in 1932–1934. On Verlaten 2 species were "lost" and 4 were "gained." This balance sheet cannot easily be dismissed as an artifact of collecting technique. Dammerman notes that during this period, "the most remarkable thing is that now for the first time true flycatchers, *Muscicapidae*, appeared on the islands, and that there were no less than four species: *Cyornis rufigastra, Gerygone modigliani, Alseonax latirostris* and *Zanthopygia narcissina*. The two last species are migratory and were therefore only accidental visitors, but the sudden appearance of the *Cyornis* species in great numbers is noteworthy. These birds, first observed in May 1929, had already colonized three islands and may now be called common there. Moreover, the *Gerygone*, unmistakable from his gentle note and common along the coast and in the mangrove forest, is certainly a new acquisition." Extinctions are less susceptible of proof, but the following evidence is suggestive: "On the other hand two species mentioned by Jacobson (1908) were not found in 1921 and have not been observed since, namely the small kingfisher *Alcedo coerulescens* and the familiar bulbul *Pycnonotus aurigaster*." Between 1919–1921 and 1932–1934 the conspicuous *Demiegretta s. sacra* and *Accipiter* sp. were "lost," although these species may not have been truly established as breeding populations. But "the well-known grey-backed shrike (*Lanius schach bentet*), a bird conspicuous in the open field, recorded in 1908 and found brooding in 1919, was not seen in 1933. Whether the species had really completely disappeared or only diminished so much in numbers that it was not noticed, the future must show." Of course the change in habitat as the island became forested is responsible for some of this turnover, at least in the early decades. Future research on the Krakatau fauna would indeed be of great interest, in view of the very dynamic equilibrium suggested by the model we have presented. If the "losses" in the data represent true extinctions, the rate of extinction would be 0.2 to 0.4 species per year, approaching the predicted rate of 0.8 to 1.6. This must be regarded as a minimum figure, since it is likely that species could easily be lost and regained all in one 12-year period.

It is also of interest to view the matter in a different way, using the actual census data as an estimate (admittedly an underestimate) of the annual extinction losses, X, *at equilibrium*. From this we can estimate t from the size of the saturated fauna, S. Using Equation (3-15) again, and substituting 0.3 for X,

$$t_{0.90} \sim \frac{2.3}{2} \times \frac{30}{0.3} = 115 \text{ years}$$

or, using Mayr's estimate of a fauna of about 42 species

$$t_{0.90} \sim \frac{2.3}{2} \times \frac{42}{0.3} = 161 \text{ years}$$

and the time, $t_{\frac{2}{3}}$, to two-thirds saturation should be very roughly

$$t_{\frac{2}{3}} \sim \frac{1.1}{2} \times \frac{42}{0.3} = 77 \text{ years,}$$

which is much closer to the actual elapsed time since the explosion.

Such might be the situation in the early history of the equilibrium fauna. It is not possible to predict whether the rate of turnover would change through time. As other taxa reach saturation, and more species of birds have a chance at colonization, it is conceivable that more "harmonic" species systems would accumulate within which the turnover rate would decline. Yet, as Lack (1942) has demonstrated for the bird fauna of the Orkney Islands, species extinctions and replacements could continue at a rate high enough to be measured within a few decades. This evolutionary aspect will be treated more carefully in Chapter 7.

When we turn to the data on the buildup of the Krakatau floras, assembled by Docters van Leeuwen, a very different story unfolds. As seen in Figures 22A and B, the numbers of plant species increased more or less steadily through the observation period, 1883–1934, and the colonization curves consequently fluctuated but did not drop. Although the 1934 expedition certainly was more thorough than the previous ones, and was the first to examine the high point of the island, the continued increase was so great as to be due to more than simply improved sampling. This result can be explained only if the immigration and extinction curves remained approximately parallel through 1934. There are two conceivable causes for such an unexpected

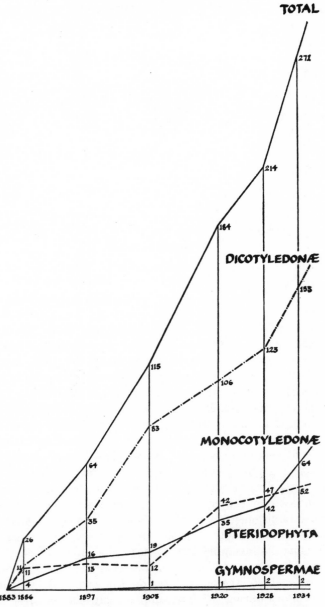

FIGURE 22A. Buildup of numbers of species of three higher plant groups on the three islands of the Krakatau group. (From Docters van Leeuwen, 1936.)

49

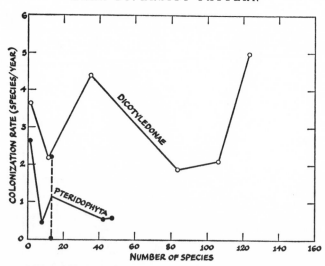

FIGURE 22B. The colonization curves for Dicotyledonae and Pteridophyta on the three Krakatau Islands. (Based on Docters van Leeuwen, 1936.)

phenomenon: either the pool of plant species on Java and Sumatra is so enormous in comparison with the number of species settled on the Krakatau islands by 1934 that a depletion effect is not visible, or else the extinction curve actually *declined* with species buildup. The flora of the greater Sunda islands is certainly enormous enough to mask a depletion effect, although the pool of effective colonizers among the plant species may not be. The second explanation seems equally reasonable. When plants as a whole are colonizing a barren area, the extinction curve should decline at first, due to succession. Later plant communities are dependent on earlier, pioneer communities for their successful establishment. Yet when they do become established, they do not wholly extirpate the pioneer communities, at least not if the sample area is topographically varied enough; and a large part of the total successional diversity is thereby preserved. Furthermore, trees do not grow tall for many years, and during this interval no deeply shaded forest can be found. The effects are illustrated in Figure 23. An ini-

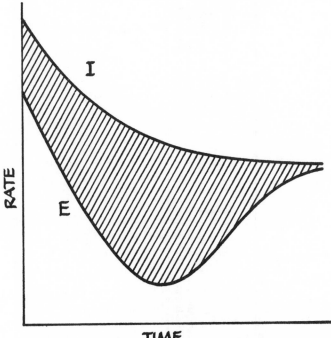

TIME

FIGURE 23. If succession is important enough in the early stages of colonization, the time curve of the extinction rate (*E*) can decline at the beginning and then rise to the equilibrium level, rather than rise monotonically throughout. The result can be a linear or even upward-curving colonization curve, such as that observed in the Krakatau flora. See text for further information.

tially declining extinction curve can also be hypothesized in cases where there is a succession involving different trophic levels. It could have happened, for example, in the earlier phase of colonization by birds on Krakatau, as perhaps suggested by the more rapid increase between 1908 and 1920 than between 1883 and 1908.

Now let us consider some evidence relating to natural extinction in floras of islands at or near the equilibrial number of species. An unusual opportunity to this end has been presented by published floristic maps of the Florida Keys. In 1904 O. E. Lansing surveyed the small islands

51

TABLE 5. Extinction of plant species in the Dry Tortugas, 1904–1916

Island	Estimated area in sq. ft. (1916)	Total no. of species (1904)	Total no. of spp. found in 1904 that were absent in 1916	Estimated % of all species extinguished during 1904–1916	No. of local spp. (1904)	No. of local spp. found in 1904 that were absent in 1916	Estimated % of local spp. extinguished during 1904–1916
East Key	1,200,000	8	0	0	3	0	0
Sand Key (Hospital Key)	36,000	5	5	100	—	—	—
Garden Key	720,000	35	6	17	24	6	25
Bird Key	150,000	12	1	8	6	1	17
Loggerhead Key	2,100,000	21	1	5	12	1	8
Middle Key	4,000	1	1	100	—	—	—

stretching from Key West to the Dry Tortugas. His collections and vegetation maps were analyzed by Millspaugh (1907). Made over a period of only 11 days, the Lansing maps were probably not complete, but they were thorough enough to permit comparison with later surveys and, more importantly, evaluation of species extinction rates. The islands were surveyed again and in much greater detail in 1916 by Bowman (1918). A comparison of the maps made in 1904 and 1916 shows some striking vegetation changes, including some species extinctions. That rapid change is a continuing process has been confirmed by a later study of Davis (1942) and the current surveys of W. B. Robertson, Jr., the Park Biologist of the Everglades National Park (personal communication).

Extinction rates in the Dry Tortugas in the period 1904–16 have been estimated by comparison of the Lansing and Millspaugh maps. The estimates are given in Table 5 and certain correlations with island area and species numbers represented in Figures 24 and 25. Since the Bowman survey was complete or nearly so, species located by Lansing but not by Bowman are to be regarded as probably having been

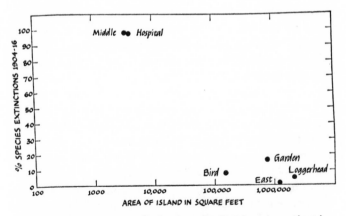

FIGURE 24. Percentage of plant species that became extinct in various islands of the Dry Tortugas between 1904 and 1916, plotted against island area. The smallest keys, Hospital and Middle, were wholly denuded by wave action. Extinctions in the larger keys were limited to local, annual species.

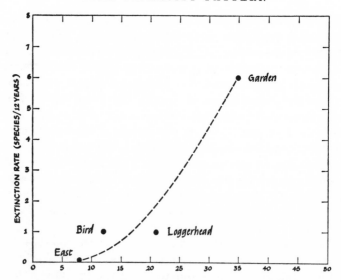

FIGURE 25. Percentage of plant species that became extinct in several Dry Tortugas keys of roughly comparable areas, as a function of species numbers (abscissa). These few data insofar as they go support the postulate that relative extinction rate increases with species density.

extinguished in the 12-year period. This conclusion is supported by the fact that Bowman's data reveal a seemingly consistent vegetational change at the extinction sites, to be discussed shortly. The estimates are to be taken as minimal, for two reasons: Lansing could have missed some rare species that became extinct, especially those that existed only as evanescent propagules; and, less likely, some species could have become extinct but have re-established themselves all within the 12-year period. We have noted only extinctions that were apparently "natural"; several local species on Garden Key that were clearly obliterated in 1904–16 by construction activity around Fort Jefferson are not included.

Several conclusions can be drawn about the extinction process in the Dry Tortugas flora. It can be seen that in the two very small islands, with areas of less than 10,000 square feet, extinction was total. This is because the islands themselves are unstable, being little more than shifting sand bars. Strand species became established easily, but the average life expectancy of their populations is only a few years—a condition that has continued to the present time (Robertson, personal communication). For these smallest populations, ecological adaptation has little to do with survivorship; *all* are extirpated periodically. In order to survive on a system of such islets where no nearby larger islands exist to serve as refugia, it would therefore be necessary for species to have an exceptionally high dispersal ability.

For the somewhat longer islands that were stable in 1904–16, extinction rates varied. In three out of four, some extinction occurred and was on the order of 10%. From Figure 25, which includes data from the four larger keys, a species-density dependence effect is apparent. East Key, an isolated island with only eight species, suffered no extinctions (and no additions). At the other extreme Garden Key, whose flora had been greatly built up through accidental introductions by man, had an extinction rate of at least 17%.

The extinctions included in these calculations probably were natural. All of the species that became extinct in the Dry Tortugas were local in 1904; they had been recorded by Lansing from only one or (occasionally) two sites. On the other hand all of the many widespread species recorded by Lansing were also found by Bowman in 1916. This evidence bolsters the intuitive notion that extinction is more likely to occur in species with small, local populations. All of the extinguished species were herbs and grasses; no shrubs or trees became extinct. Extinction of these short-lived forms was usually associated with encroachment of the site by shrubs or trees, suggesting that competition of the kind associated with succession was responsible. No species became extinct on more than one island. As a consequence, we can conclude that species that can establish populations on more than one larger island in an archipelago such as the Dry Tortugas greatly reduce the risk of extinction.

Finally, both colonization and extinction of fresh-water diatoms have recently been measured in an ingenious set of experiments by Ruth Patrick, who has kindly permitted us to include the following brief account prior to the publication of her own report. Dr. Patrick suspended glass slides of different surface areas in Roxborough Spring in Pennsylvania and counted and identified the diatoms which were attached after one week and, in a replicate, after two weeks. Hence she secured species counts on "islands" of small and large size one and two weeks after these islands were opened for colonization. Four actual censuses are reproduced here (Table 6), since they show not only that the large area holds more species, but also that competition, which ensued as the diatoms became crowded, reduced the numbers of species. Furthermore, it was generally the rare species that became extinct. Thus in at least three features—the dependence of rate on species density, the role of competition, and the differential elimination of rare species—the extinction process in these microorganisms paralleled that in the higher plants of the Dry Tortugas.

Faunal Differences Among Islands and the Shape of the Immigration Curve

Knowledge of the *number* of species on islands and variance of the number among islands of the same area can provide an idea of the degree to which equilibrium has been approached. So far extra knowledge of the *names* of the species has been wasted. It can be put to effective use, however, in calculating the shape of the immigration curve.

Since calculation of the immigration curve is complex, we will approach it here in stages. We start with the extreme case in which every species in the source region is equally likely to immigrate and persist. Then the immigration curve falls by a constant amount between each two successive species. If there are P species available for immigration, the immigration curve begins at height I_0 at the left and falls linearly to zero at P species, as shown in Figure 26. In the case of such an immigration curve, two islands that derive their biotas from the same sources with the same rates of immigration and extinction may themselves have quite different sets of species. In fact, since all species are equally probable, the $\hat{S}$ species on the first island are a random sample of the P available, and the probability is $\hat{S}/P$

TABLE 6. Censuses of diatom species settled on glass slides, of differing areas, which were suspended in Roxborough Spring, Pennsylvania. Counts were made at the ends of the first and second weeks. (Unpublished data of Ruth Patrick.)

Exposure time: Slide area:	Experiment 1		Experiment 2		Experiment 3		Experiment 4	
	One week 12 mm²	Two weeks 12 mm²	One week 25 mm²	Two weeks 25 mm²	One week 12 mm²	Two weeks 12 mm²	One week 25 mm²	Two weeks 25 mm²
Achnanthes lanceolata var. *elliptica*	60	261	226	810	4	40	126	180
Achnanthes lanceolata	1,592	11,310	2,842	170,100	120	3,132	2,610	61,902
Achnanthes minutissima	150	320	805	540	8	20	487	120
Amphora ovalis var. *pediculus*	966	931	4,717	8,730	84	175	4,066	240
Diatoma hiemale var. *mesodon*	4,011	33,400	7,912	128,800	220	1,827	11,250	7,783
Diploneis marginestriata	192	348	3,168	2,460	50	80	2,942	412
Eunotia pectinalis	2	70	63	2,250	3	23	36	350
Fragilaria pinnata	66	180	1,422	4,140	72	104	944	540
Gomphonema longiceps var. *subclavata*	84	680	272	3,960	4	95	486	1,086
Gomphonema parvulum	1,114	51,940	4,563	118,080	158	7,917	4,968	14,392
Melosira islandica fo. *curvata* cf. *spiralis*	1		18		1		2	
Meridion circulare var. *constricta*	726	1,914	540	11,520	22	120	306	1,448
Navicula atomus	36	60	172	280	6	7	144	75
Navicula cryptocephala	42	30	389	300	18		378	20

TABLE 6. (Continued)

Exposure time: Slide area:	Experiment 1 One week 12 mm²	Two weeks 12 mm²	Experiment 2 One week 25 mm²	Two weeks 25 mm²	Experiment 3 One week 12 mm²	Two weeks 12 mm²	Experiment 4 One week 25 mm²	Two weeks 25 mm²
Navicula pelliculosa	2		6					
Navicula seminuloides	120	90	570	2,430	22	140	324	1,267
Navicula seminulum	60	6	190	360	5	3	180	15
Nitzschia amphibia	5		2	90			14	20
Nitzschia dissipata	4		8					
Nitzschia frustulum perminuta	5	60	90	180	2	5	26	105
Nitzschia palea var. debilis	5	4	9	18	2		11	4
Nitzschia tryblionella var. debilis	7	6	290	990	6	8	162	35
Opephora martyi	8		62	20	3		18	2
Pinnularia gibba var. mesogongyla	1	1	45	40		2	47	3
Stauroneis pygmaea	4		36	14	2		18	4
Synedra rumpens	160	140	190	9,090	20	85	342	1,086
Navicula placenta	1		3					
Navicula minuscula	9	12	235	360	4		108	70
Navicula vitabunda	12	40	326	630	16		360	362
Navicula hungarica var. capitata	2		3	1			9	

58

TABLE 6. (*Continued*)

	Experiment 1		Experiment 2		Experiment 3		Experiment 4	
Exposure time:	One week	Two weeks	One week	Two weeks	One week	Two weeks	One week	Two weeks
Slide area:	12 mm²	12 mm²	25 mm²	25 mm²	12 mm²	12 mm²	25 mm²	25 mm²
Fragilaria pinnata var.	1		4	10	1	150		4
Navicula bulnheimii	2		27		4		46	
Navicula contenta forma biceps			14		1		10	
Navicula near paratunkae			61				36	
Navicula pupula			6					
Navicula similis			57					
Navicula tantula			35				18	
Nitzschia paleacea			38				2	
Nitzschia near communis			1					
Pinnularia microstauron		60	135	450	9	60	126	310
Pinnularia mesolepta var. *stauroneiformis*			9				4	
Cymbella aspera			1				2	
Cyclotella melosiroides			34				1	
Diploneis puella			1					
Neidium iridis var. *amphigomphus*			3					
Synedra parasitica		1	4	12			14	2
Synedra vaucheriae var. *capitellata*			2					

TABLE 6. (*Continued*)

	Experiment 1		Experiment 2		Experiment 3		Experiment 4	
Exposure time:	One week	Two weeks	One week	Two weeks	One week	Two weeks	One week	Two weeks
Slide area:	12 mm²	12 mm²	25 mm²	25 mm²	12 mm²	12 mm²	25 mm²	25 mm²
Stauroneis phoenicenteron							10	
Navicula mutica							1	
Navicula mutica var.							2	
Hantzschia amphioxys							1	
Synedra ulna (Nitzsch.) Ehr.							1	
Frustulia vulgaris							3	
Nitzschia linearis						1	2	
TOTALS	9,450	1,864 (100,000)	29,600	466,665	8,670	13,994	30,643	91,837
No. of taxa	32	23	47	29	28	21	44	28

60

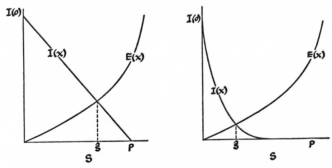

FIGURE 26. *Left:* The linear immigration curve, with intersections at $I(0)$ and P, which will result in the extreme case where the immigration rates are equal for all species in an immigrant pool containing P species. $I(x)$, immigration rate in species/unit time. S, number of species already arrived on the island. $\hat{S}$, equilibrial number on the island. *Right:* The exponential curve resulting in the opposite extreme case where the immigration rates of different species are so different that islands will fill up by a highly regular progression beginning with the species with the highest rate and ending with the one with the lowest rate.

that any particular one of these will also be found on the second island. Thus a proportion $1 - \hat{S}/P$ of the species on the second island will be different, so that the second island contains $\hat{S}(1 - \hat{S}/P)$ species not found on the first. Hence the two islands combined will contain $\hat{S} + \hat{S}(1 - \hat{S}/P) = \hat{S}(2 - \hat{S}/P)$ species.

Now working backward, it can be seen that if the difference in species composition for a given taxon is this great, that is, if the proportion of species *not* held in common is $(P - \hat{S})/(2P - \hat{S})$, then the immigration curves are linear. In making any such evaluation, care must be taken because differences in species composition will also arise if the islands are ecologically very different. We have also assumed that subsequent successful new immigrant species will not replace the earlier ones, at least not in any regular fashion.

The opposite extreme case from linearity is the one in which the species in the immigrant pool can be ranked from the most likely immigrant to the least, and each species is sufficiently more probable than all the lesser-ranked ones combined that it is very likely to precede them onto the island. The shape of this extreme immigration curve can be roughly calculated as follows. The drop in the immigration curve $I(x)$ between x species and $x + 1$ species on the abscissa is proportional to the probability of the $(x + 1)^{\text{th}}$ species arriving on the island. This, we have said, must be a suitably large multiple, say 9, of the accumulated probability of all the remaining species. In symbols,

$$-\frac{dI(x)}{dx} = c \int_{x}^{P} I(x)\, dx,$$

whence

$$\frac{d^2I}{dx^2} - cI(x) = 0.$$

Trying e^{mx} as a solution, we find it works if $m^2 = c$, i.e., $m = \pm \sqrt{c}$. Therefore,

$$I(x) = Ie^{-\sqrt{c}x},$$

(with c a constant value about 9 to approach the usual level of statistical confidence), as can be verified by substitution. This is the immigration curve, and it is very concave (Figure 26). In order for the species to arrive with relative certainty in the ranked order from most to least probable, the immigration curve must be at least this concave. Hence, if two similar islands have the same species, and these species were drawn from a relatively large pool, the immigration curves must be as concave as shown in Figure 26.

Of course most pairs of real islands which are otherwise similar will have faunal differences that lie between these two extremes. Consequently their immigration curves will be in some respects intermediate. Roughly, then, the number of species on two islands will lie some fraction of the way between the number on a single island and $2 - (\hat{S}/P)$ times that number, and the immigration curve can be thought to be approximately that same fraction of the way between the sharply bending curve of certain order and the flat immigration curve.

The use of this formula is suggestive at best. It should not be applied to islands that are ecologically very dissimilar or to islands so small that new immigrant species always have a good chance of replacing existing ones. Neither should it be applied to islands so remote that they develop considerable numbers of separate endemic species.

An example of a fauna whose immigration curve lies between the two extremes is given in Table 7. There are

TABLE 7. Distribution of ant species on ten very small mangrove islands in the lower Florida Keys. (D. Simberloff and E. O. Wilson, unpublished data.)

Species pool	Code number of island									
	1	2	3	4	5	6	7	8	9	10
Pseudomyrmex elongatus	×	×	×	×	×	×	×	—	×	×
Pseudomyrmex pallidus	—	—	—	—	—	—	—	×	—	—
Crematogaster ashmeadi	×	×	×	×	×	×	×	×	×	×
Xenomyrmex floridanus	—	—	×	—	—	×	—	×	×	×
Monomorium floricola	—	—	—	—	—	×	—	×	—	—
Paracryptocerus varians	×	×	×	×	—	×	×	×	—	×
Tapinoma littorale	—	×	×	×	—	×	×	—	×	—
Paratrechina bourbonica	—	—	×	—	—	—	—	—	—	—
Camponotus floridanus	—	—	—	—	—	—	—	—	—	—
Camponotus tortuganus	—	—	—	—	—	×	—	—	—	—
Camponotus impressus aff.	—	×	×	—	×	—	×	—	×	—

eleven arboricolous ant species found commonly in the very small red mangrove islands of the Florida Keys. The ten islands catalogued here each have an area of about 1,000 square feet and hold between three and seven species, with an average of five species per island. Even a casual inspection of the data will show that the faunulae of individual islands is by no means a random sample of the pool of eleven species. The tendency of the faunulae to form nested subsets of each other is marked, suggesting wide variation among the species in dispersal power. On the other hand, this nested pattern of distribution, from widespread to less widespread species, is far from perfectly regular. This proves on the one hand that the species' dispersal powers are not widely different enough to ensure a completely predictable sequence of colonists, and on the other hand that colonizing species do not extirpate earlier colonists, at least not with regularity and speed.

Weaknesses of the Equilibrium Model

The equilibrium model has the virtues of making testable predictions which were not immediately obvious and of making the individual vagaries of island history seem somewhat less important in understanding the diversity of the island's species. Of course the history of islands remains crucial to the understanding of the taxonomic composition of species. The equilibrium model further makes a valid distinction between islands which are early and late in their colonization. It has at least three weaknesses, however. First, we still know very little about the precise shape of the extinction and immigration curves, so that few numerical predictions can yet be made. Second, and more importantly, the model puts rather too simple an interpretation on the process by making an artificially clearcut distinction between immigration and extinction. Actually, the two phenomena are often difficult to identify. For instance, is a migrating bird, passing through an island, called an immigrant? If so, it goes extinct as it leaves. Suppose we agree to ignore species not "intending" to stay. But what about a solitary male which arrives on the shore perfectly willing to colonize, but without a mate? Again, if we call this an immigrant, we are sure of extinction shortly. There is nothing wrong with this convention except that it magnifies "extinction" beyond what most persons would accept. We could ignore immigrants that do not colonize with the potential of reproducing. Then both immigration and extinction would be enormously reduced, but even most of these colonies are doomed in their early years; and these failures to succeed at colonization undoubtedly account for most of the extinctions. This is what we prefer to call immigration and extinction, but we emphasize that this preference is quite arbitrary. It would be possible to be still more conservative and not even count colonies that fail as immigrants. Only "successful" colonists would be immigrants, and extinction would then be an exceedingly rare event. The trouble with this view is that extinction during the early stages of colonization is not always easy to distinguish from the extinction of an established colony. In other words, it is not easy to say when a colony is established. As we shall see later (in Chapter 4), when a colony can grow to a large size, then failure during initial settlement is quite distinguishable from failure after the population has settled successfully; but when the population is restricted by shortage of habitat, then this distinction is no longer clear.

The third difficulty stems from the assumption that the extinction and immigration curves have a fairly regular shape for different faunas and different islands and for different times on the same island. When a new set of curves must be derived for a new situation, the model loses much of its virtue. Deviant cases do occur—for example the Krakatau flora. The extinction curves, furthermore, have a pronounced genetic component: rarity affects gene frequencies, and genetic as well as ecological causes of extinction act in concert.

These considerations should make it clear that the important aspects of the equilibrium model can be more profitably extended by a detailed consideration of the course of new colony foundation. This we will undertake in the next chapter.

SUMMARY

Multiple regression analyses have shown that area alone accounts for most of the variation in species numbers on islands. But area itself is correlated with environmental diversity, which exerts a more direct effect on species numbers and is a quality that has only begun to be described and measured. As regression studies are pressed in the future, it should be possible gradually to eliminate components of environmental diversity that are irrelevant. At the same time, only equilibrium models are likely to lead to new knowledge concerning the dynamics of immigration and extinction. A biotic equilibrium is said to be reached in a taxon on a given island when the immigration and extinction rates, measured in species/unit time, equal each other.

In this chapter, a basic equilibrium model is developed which postulates immigration rate curves that fall and extinction rate curves that rise with an increase in the number of resident species. The model leads to the prediction that the logarithm of the number of species should increase with area more rapidly on distant islands than on near islands, and should decrease more rapidly with distance on small islands than on large islands. The limited evidence has so far proved consistent with this prediction. Other alterations in the model parameters, such as clumping of the islands, addition of stepping stones, and suspension of density dependence due to rapid species turnover on very small islands, result in characteristic alterations of the area-species curve.

A stochastic model of immigration and extinction on single islands has also been developed, leading to the following approximate equation:

$$X = \frac{1.16\hat{S}}{t_{0.90}}$$

where X is the extinction rate at equilibrium (therefore, the immigration rate as well), $\hat{S}$ is the mean number of species at equilibrium on islands of a given kind, and $t_{0.90}$ is the time required to reach 90% of the equilibrial number of species. The equation allows a surprisingly high turnover rate in species for islands that are colonized quickly. But, again, the limited available evidence seems consistent.

The measurement of immigration and extinction rates in nature will be difficult to accomplish. There are two ways of approaching the problem indirectly: (1) by measuring the variation in species numbers on similar islands and working back to immigration and extinction rates by means of the stochastic model for single islands; and (2) by using the variation in species composition among similar islands and deducing the shape of the immigration curve.

Colonization curves, on the other hand, are relatively easy to obtain. Several are already available, and they present some interesting and intriguing results. The colonization curve drawn as a function of time is the integral through time of the difference between the immigration and extinction time curves. If the colonization curve is known and either the immigration or extinction curve can be estimated or measured directly, then the third curve can be drawn, and the whole system thereby specified. This remains to be achieved in any single case.

CHAPTER FOUR

The Strategy of Colonization

In this chapter we shall attempt to relate the properties of the life history of a colonizing species to its chances for success and, if it fails, to the length of time it persists before going extinct. There is only a small amount of theory already available on this problem and a considerable volume of empirical but rather unsystematic lore, much of which has been collected in the recent volume on colonizing species edited by Baker and Stebbins (1965). Here we shall summarize and extend the theory and retell some of the lore most relevant to it.

Theory

We first define "birth" and "death" probabilities, λ_x and μ_x,[1] as follows. Let the birth rate of the whole population per unit time be λ_x so that the probability of the population changing from size x to $x + 1$ in a small time interval, h, is $\lambda_x h$. If h is suitably small, the probability of increases from x to $x + 2$ or more is negligible. Let μ_x be the mortality rate of the whole population so that the probability of change from x to $x - 1$ in the interval h is $\mu_x h$. λ_x and μ_x are allowed to vary with x in different ways to account for any pattern of density dependence.

Just two properties of λ_x and μ_x are needed. First, the probability of *any* population change from x in the interval of length h is $(\lambda_x + \mu_x)h$. (The probability of both increase and decrease in the interval is negligibly small.) The probability of an increase in the interval is $\lambda_x h$, so that the fraction of changes that are increases is given by $\lambda_x/(\lambda_x + \mu_x)$. Similarly the fraction of changes that are decreases is given

---

[1] λ and μ in this chapter refer to birth and death rates of organisms within a single population. In Chapter 3 the same symbols were used, with different subscripts, to designate the analogous measures of immigration and extinction of whole populations.

by $\mu_x/(\lambda_x + \mu_x)$. The second required property is the expected (average) time per change. Since the expected number of changes per unit time is $(\lambda_x + \mu_x)$, the expected time per change is $1/(\lambda_x + \mu_x)$.[1]

We will now construct a model that can be used to estimate the mean survival time of populations. This is the most relevant parameter of colonizing success. It would be feasible to begin the analysis instead with an estimate of the probability that a given population will become extinct in any given period of time. But since all populations are limited in their maximum size by the carrying capacity of the environment (given as K individuals), and since all individuals have a chance of dying, all populations will surely go extinct if given enough time. Of more interest is the fact that soon after colonizing the island, and while its numbers are still very low, a given population will have a short expectation of life; but later, if it survives and reaches a large size, it will normally persist for an enormous time. Hence it is better to consider directly the expected time before the inevitable extinction.

(The reader who does not wish to consider the mathematical argument in detail at this time can pass over the following section taking up instead the principal conclusions that begin after the next rule.)

---

[1] More formally, the probability of no change in time interval $t + h$, $[P_0(t + h)]$, is related to the probability of no change in interval t as follows. $P_0(t + h) = P_0(t)[1 - (\lambda_x + \mu_x)h]$ if h is small. Rearranging terms,

$$\frac{dP_0(t)}{dt} = \lim_{h \to 0} \frac{P_0(t + h) - P_0(t)}{h} = -P_0(\lambda_x + \mu_x),$$

whence $P_0(t) = e^{-(\lambda_x+\mu_x)t}$. This is the probability of an interval of at least t preceding any change. The probability density of exactly t is the negative derivative:

$$(\lambda_x + \mu_x)e^{-(\lambda_x+\mu_x)t},$$

whose expected value is

$$\int_0^\infty t(\lambda_x + \mu_x)e^{-(\lambda_x+\mu_x)t}\, dt = \frac{1}{\lambda_x + \mu_x}.$$

The expected time T_x for a population of size x to go extinct, i.e., the mean survival time of the population, is the sum of the time $1/(\lambda_x + \mu_x)$ preceding any change plus the time to extinction from the position of the first change. Recall that the population goes from x to $x + 1$ with probability $\lambda_x/(\lambda_x + \mu_x)$ and from x to $x - 1$ with probability $\mu_x/(\lambda_x + \mu_x)$. In symbols,

$$(4\text{-}1) \quad T_x = \frac{1}{\lambda_x + \mu_x} + \frac{\lambda_x}{\lambda_x + \mu_x} T_{x+1} + \frac{\mu_x}{\lambda_x + \mu_x} T_{x-1},$$

and we know $T_0 = 0$. These two equations can be employed to relate the life history parameters λ and μ to the success of colonization. It is essential to build into the scheme the properties of exponential growth (on the average) for small populations, as well as a ceiling, K, beyond which the population cannot normally grow. There are any number of ways of approaching population size K and staying there at equilibrium. For simplicity we will recognize two extremes. In the first, λ_x is proportional to x ($\lambda_x = \lambda x$) until x reaches $K + 1$ whereupon it drops to 0, while μ_x is proportional to x ($\mu_x = \mu x$) for all x ("birth but not death is density dependent"). In the second extreme, we let μ_x be proportional to x until x reaches $K + 1$, whereupon μ_x becomes such that it instantly reduces the population to K, while λ_x is always proportional to x ("death, but not birth, is density dependent"). When *birth* is density dependent, the coefficients in Equation (4-1) become $1/(\lambda + \mu)x, \lambda/(\lambda + \mu)$, and $\mu/(\lambda + \mu)$ for $x \leq K$, where λ and μ are constants, the per capita instantaneous birth and death rates. For $x > K$, we substitute 0 for λ so that the coefficients are $1/\mu x$, 0, and 1, respectively. When *death* is density dependent the coefficients are $1/(\lambda + \mu)x$, $\lambda/(\lambda + \mu)$, and $\mu/(\lambda + \mu)$ for $x \leq K$ and 0, 0, 1, respectively, for $x > K$. Under both of these regimes, the population grows exponentially to level K, at which point it stops abruptly. We treat these cases as the ones most favorable to colonization. All other kinds of density dependence would give more rapid extinction. We can now solve (4-1) explicitly by writing out the equation for all x up to K. Starting with *"death density dependent,"*

$$T_1 = \boxed{\frac{\lambda}{\lambda + \mu} T_2} \qquad\qquad + \frac{1}{\lambda + \mu},$$

$$T_2 = \boxed{\frac{\lambda}{\lambda + \mu} T_3} + \frac{\mu}{\lambda + \mu} T_1 + \frac{1}{2(\lambda + \mu)},$$

$$T_3 = \frac{\lambda}{\lambda + \mu} T_4 + \boxed{\frac{\mu}{\lambda + \mu} T_2} + \frac{1}{3(\lambda + \mu)},$$

$$T_4 = \frac{\lambda}{\lambda + \mu} T_5 + \boxed{\frac{\mu}{\lambda + \mu} T_3} + \frac{1}{4(\lambda + \mu)},$$

$$\vdots$$

$$T_K = \frac{\lambda}{\lambda + \mu} T_{K+1} + \frac{\mu}{\lambda + \mu} T_{K-1} + \frac{1}{K(\lambda + \mu)}.$$

Summing these equations, and bracketing the enclosures on the right whose coefficients add to 1, we obtain

$$T_1 + T_2 + T_3 + \cdots + T_K$$
$$= \frac{1}{\lambda + \mu}\left[1 + \frac{1}{2} + \frac{1}{3} + \cdots + \frac{1}{K}\right] + \frac{\mu}{\lambda + \mu} T_1$$
$$+ T_2 + T_3 + \cdots + T_{K-1} + \frac{\lambda}{\lambda + \mu} T_K + \frac{\lambda}{\lambda + \mu} T_{K+1}.$$

Subtracting $T_2 + T_3 + \cdots + T_{K-1}$ from each side and simplifying,

$$(4\text{-}2) \quad \frac{\lambda}{\lambda + \mu} T_1 + \frac{\mu}{\lambda + \mu} T_K = \frac{\lambda}{\lambda + \mu} T_{K+1}$$
$$+ \frac{1}{\lambda + \mu}\left[1 + \frac{1}{2} + \frac{1}{3} + \cdots + \frac{1}{K}\right].$$

Finally, notice that

$$T_{K+1} = 0 \cdot T_{K+2} + 1 \cdot T_K + 0$$

so that

$$T_{K+1} = T_K.$$

Substituting this and cancelling $1/(\lambda + \mu)$,

$$\lambda T_1 = (\lambda - \mu) T_K + \left[1 + \frac{1}{2} + \frac{1}{3} + \cdots + \frac{1}{K}\right]$$

which leads to

$$(4\text{-}3) \quad T_K = \frac{\lambda}{\lambda - \mu} T_1 - \frac{1}{\lambda - \mu} \left[1 + \frac{1}{2} + \cdots + \frac{1}{K} \right],$$

when "deaths are density dependent."

Starting with *"birth density dependent,"* Equation (4-1) produces

$$T_{K+1} = 0 \cdot T_{K+2} + 1 \cdot T_K + \frac{1}{\mu(K+1)}$$

$$= T_K + \frac{1}{\mu(K+1)}.$$

Substituting this in Equation (4-2) leads to

$$\lambda T_1 = (\lambda - \mu) T_K + \frac{\lambda}{\mu(K+1)}$$
$$+ \left[1 + \frac{1}{2} + \frac{1}{3} + \cdots + \frac{1}{K} \right],$$

whence

$$(4\text{-}4) \quad T_K = \frac{\lambda}{\lambda - \mu} T_1 + \frac{\lambda}{\mu(K+1)(\lambda - \mu)}$$
$$- \frac{1}{\lambda - \mu} \left[1 + \frac{1}{2} + \frac{1}{3} + \cdots + \frac{1}{K} \right]$$

when "births are density dependent."

Equations (4-3) and (4-4), relating T_K to T_1, are half of what we need. We next find T_1 by rewriting Equation (4-1), obtaining successively

$$T_2 - T_1 = \frac{\mu}{\lambda}(T_1 - T_0) - \frac{1}{\lambda} = \frac{\mu}{\lambda} T_1 - \frac{1}{\lambda},$$

$$T_3 - T_2 = \frac{\mu}{\lambda}(T_2 - T_1) - \frac{1}{2\lambda} = \left(\frac{\mu}{\lambda}\right)^2 T_1 - \frac{\mu}{\lambda} \cdot \frac{1}{\lambda} - \frac{1}{2\lambda},$$

$$T_4 - T_3 = \frac{\mu}{\lambda}(T_3 - T_2) - \frac{1}{3\lambda} - \frac{1}{K\lambda} = \left(\frac{\mu}{\lambda}\right)^3 T_1$$
$$- \left(\frac{\mu}{\lambda}\right)^2 \cdot \frac{1}{\lambda} - \frac{\mu}{\lambda} \cdot \frac{1}{2\lambda} - \frac{1}{3\lambda},$$

$$T_{K+1} - T_K = \frac{\mu}{\lambda}(T_K - T_{K-1}) - \frac{1}{K\lambda} = \left(\frac{\mu}{\lambda}\right)^K T_1$$
$$- \left(\frac{\mu}{\lambda}\right)^{K-1} \cdot \frac{1}{\lambda} - \left(\frac{\mu}{\lambda}\right)^{K-2} \cdot \frac{1}{2\lambda} - \cdots - \frac{1}{K\lambda}.$$

We could add x of these to get a cumbersome formula for T_x in terms of T_1, but all we need here is to notice that $T_K = T_{K+1}$ when deaths are density dependent (*death density dependent*) so that from the last equation,

$$(4\text{-}5) \quad T_1 = \frac{\lambda}{\mu} \cdot \frac{1}{\lambda} + \left(\frac{\lambda}{\mu}\right)^2 \cdot \frac{1}{2\lambda} + \left(\frac{\lambda}{\mu}\right)^3 \cdot \frac{1}{3\lambda}$$
$$+ \cdots + \left(\frac{\lambda}{\mu}\right)^K \cdot \frac{1}{K\lambda}.$$

Similarly if births are density dependent, so that $T_{K+1} - T_K = 1/\mu(K+1)$, the last equation in (4-4) yields (*birth density dependent*):

$$(4\text{-}6) \quad T_1 = \frac{\lambda}{\mu} \cdot \frac{1}{\lambda} + \left(\frac{\lambda}{\mu}\right)^2 \cdot \frac{1}{2\lambda} + \left(\frac{\lambda}{\mu}\right)^3 \cdot \frac{1}{3\lambda}$$
$$+ \cdots + \left(\frac{\lambda}{\mu}\right)^K \cdot \frac{1}{K\lambda} + \left(\frac{\lambda}{\mu}\right)^K \cdot \frac{1}{\mu(K+1)}.$$

We are in a position neither to know λ, μ, and K for most island species, nor to test the equations by adducing evidence except in a few restricted cases. Consequently the primary interest in the equations must be qualitative. Even at this level, however, some striking and perhaps unexpected relations among the parameters can be discovered. Figures 27A, 27B, and 27C show how T_1 changes with a variety of values of λ, μ, and K. Notice that as K increases there is a rather sharp increase in T_1 (as well as T_K), so that for small values of K, T_1 is short, while for increasing values of K it soon becomes prodigiously long.

To form an idea of the magnitudes for T_1 and T_K, we note first that for $\lambda < \mu$, all terms with λ/μ raised to some power are less than those terms with the λ/μ replaced by one. Hence, for $\lambda < \mu$ and *death density dependent*:

$$(4\text{-}7) \quad T_1 < \frac{1}{\lambda}\left[1 + \frac{1}{2} + \frac{1}{3} + \cdots + \frac{1}{K}\right]$$
$$\sim \frac{1}{\lambda} \ln (2K - 1).$$

Even for $K = 10,000$ so that $2K - 1$ is about e^{10} we learn

73

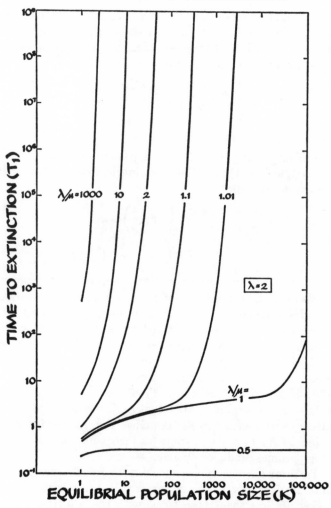

FIGURE 27A. T_1, the expected (mean) survival time of a population beginning with a single propagule, as a function of the per capita birth rate λ, per capita death rate μ, and maximum number K of individuals belonging to the species that the island can hold, where $\lambda = 2$. The propagule is defined as the minimal number of individuals capable of reproducing—ordinarily a gravid female, a seed, or an unmated female plus male. These curves are based on the condition of density-dependent death discussed in the text.

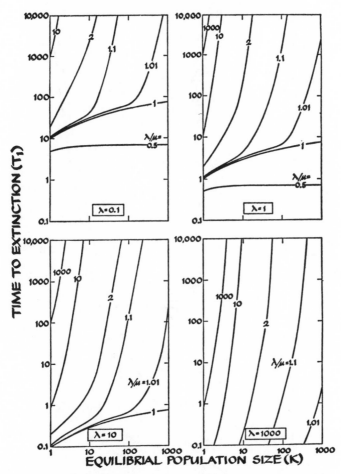

FIGURE 27B. Families of curves similar to that in Figure 27A but with differing values of the per capita birth rate λ.

that with a single propagule landing on the islands the species can expect to last less than $10/\lambda$ years, where λ is measured in per capita reproduction per year. In other words T_1 is very brief.

The more interesting case occurs when $\lambda > \mu$. For large K the value of T_1 is immense. From Figure 27A, if $K = 100$ (a very small carrying capacity), $\lambda = 2/\text{yr}$, $\mu = 1.82/\text{yr}$,

75

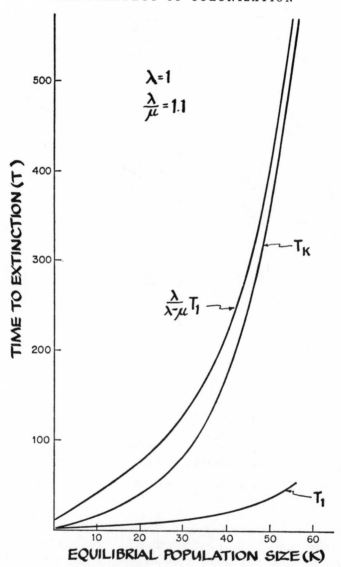

FIGURE 27C. Both T_1 and T_K are shown for $\lambda = 1$, $\lambda/\mu = 1.1$. The approximation $T_1\lambda/(\lambda - \mu)$ for T_K is also shown to indicate the degree of accuracy.

$\lambda/\mu = 1.1$, and $\lambda - \mu = r = 0.18$, we find T_1 is approximately 860 years while for $K = 1,000$ it has jumped enormously to about 10^{41} years and the expected duration of a unit propagule is stupendous—far longer than the age of the universe! This does not mean, however, that all unit propagules will persist that long. Equation (4-3) (or 4-4) gives us the means to explore this, because these say that T_K is roughly $\lambda/(\lambda - \mu)$ times as great as T_1. (The negative term is negligible for large K.) Now the fate of a propagule with $\lambda > \mu$ is either rapid extinction or a rise to a population size near K from which extinction takes extremely long. Note that if extinction either takes a very long time, T_K, or else a time so short that relative to T_K it can be taken as zero, then the mean time, T_1, for a propagule to go extinct is T_K times the probability of reaching size K. But if

$$T_K \sim \frac{\lambda}{\lambda - \mu} T_1 \text{ or } T_1 \sim \frac{\lambda - \mu}{\lambda} T_K,$$

then we can say that about a fraction $(\lambda - \mu)/\lambda$ of the propagules of size 1 will reach size K and take time T_K to go extinct, while the remaining fraction $1 - (\lambda - \mu)/\lambda = \mu/\lambda$ will go extinct very quickly. The reasoning is approximate but breaks down only if K is so small that early and late extinctions cannot be separated (see Figure 27C).

Hence we have arrived at two predictions of some interest:

(a) The chance of a single pair reaching a population near size K and taking time T_K to go extinct is about $(\lambda - \mu)/\lambda$, while that of going extinct very rapidly is about μ/λ.

(b) The time, T_K, that an established population takes to go extinct averages about $\lambda/(\lambda - \mu)T_1 = (\lambda/r)T_1$, and we can calculate it from Figure 27.

Looked at another way, immigrants must come in about once every T_1 years to maintain a species. Thus the immigration rate of a species must be at least $1/T_1$ and, conversely, the number of species on the island can grow until the values of K and λ (which decrease with the addition of new species) are such that

$$T_1 = 1/\text{immigration rate.}$$

Discussion of Survivorship Theory

We have seen that chances for quick success of a colonizing propagule should be large when $(\lambda - \mu)/\lambda = r/\lambda$ is large, and the duration of the success is large when K is large and when r/λ is large. These are intuitively fairly obvious although how they are combined to give the probability of success is not. We turn now to a discussion of how r can be large, λ small, and K large.

We begin with r and λ. r large and λ small means the propagule should have a large value of $r = \lambda - \mu$ but accomplished by a small death rate rather than by a large birth rate. Hence successful island populations should have these characteristics. But aside from the demographic problem of having large r with small λ, there is the spatial one. If colonists disperse rapidly they will not find mates and r will not even be positive, let alone large. Hence cohesiveness of the propagule is essential. The other aspect of importance is competitive reduction in r. When no competitors are present r can take its full value. Each competitor reduces r toward zero and drastically increases the proportion $\mu/\lambda = 1 - r/\lambda$ which fail to reach a safe population size. When an island is already nearly saturated with species, then r is virtually zero and almost no propagules reach a safe size.

The K effect is equally striking. Small islands and islands with many competitors provide only small values of K and have drastically reduced persistence times for established populations. When K is small (more accurately, when $(\lambda/\mu)^K$ is small) Figures 27A–B show that the expected time to extinction is small, and colonists must reinvade frequently to maintain the species. Of course K is virtually zero if the island habitat is unsuitable for the colonists; it is also small if the suitable habitat comes in small, rather isolated patches.

While the model tells a great deal about the role of K in the duration of populations, it says nothing about the factors that determine K or the variability of K through time. If K is maintained indefinitely at some relatively high number, say 1,000 or above, the mean survival time of saturated populations will remain very high. But we know that much extinction, especially that due to storms, drought, invasion by new competitors, and destruction of the environment by man, is accompanied by a severe reduction in K. If the minimum K reached by populations during such crises is known, extinction rates can be roughly estimated by modifications of our survivorship model. But changes in K cannot be predicted by the model. Nor can changes in r, due to alterations of the environment, be predicted by the model.

The principal weakness in the internal structure of the model is the coarse form of the density-dependent controls. In essence, the populations are allowed to grow without homeostatic controls until they reach the maximum size permitted by the environment (K), then they are stopped abruptly. There are perhaps populations in nature whose growth approximates the model, but growth control of most real populations is probably more gradual and complex. While of very doubtful accuracy in the analyses of particular cases whose homeostatic controls are unknown, the model nevertheless has some general usefulness. First, it allows the first qualitative generalizations about the role of the demographic parameters in population survivorship, which we have already stressed and believe to be of wide applicability. Further, since the controls postulated are crude and would result in a more rapid approach to K than more gradual complex controls, the time to extinction estimates for populations near K probably serve as a fair upper boundary for all such estimates. This is true, not because crude controls allow greater stability near K—the opposite is the case—but only because K is attained more quickly. When there is a certain density-independent mortality rate μ, and density-dependent mortality takes up the slack, then our model gives an upper bound only when the density-independent component, μ, is used. If μ is zero, there will be no extinctions.

The above model is unrealistic in another way. It assumes that all individuals are identical and that all have a uniform chance of dying in any interval of time. If, by contrast, a very few of the individuals—which we may call spores—had a very great expectation of further life and yet were still reproductive, then the time to extinction would increase greatly. We can also picture an organism which lived in a heterogeneous environment such that in one kind of patch the organisms had very low mortality while in the other patches the mortality is high. Even if the low-mortality patches are rare, they greatly increase the expected time to extinction. As it stands, our upper bound for the expected time to extinction does not apply to such cases. However, we can modify it to be useful by using the value of μ from the low-mortality individuals and the estimate of λ from those with a high rate of reproduction. This would appear to be an overestimate—a generous one—for the expected time to extinction.

Evidence Relating to the Survivorship Model

In Table 8 are listed ecological qualities of species considered by biogeographers of three animal and plant groups to be good colonizers. These have been partitioned into the general categories of attributes which appear to us, from *a priori* considerations, to contribute to colonizing success. By itself the identification of the general categories might seem at first to reduce, as Lewontin (1965) has recently declared, "the problem of colonizing species to a trivial one. The best possible organism is easy to specify." But different species and groups of species have undoubtedly qualified in different categories, and in different ways, and therein lies the possibility of a whole new level of biological investigation. For example, as Wollaston (1877) long ago pointed out, the successful beetle colonists of Saint Helena were mostly wood borers (cossonine weevils) or forms adapted to clinging to vegetation and bark (anthribid weevils); these characteristics must have given them an advantage in long-distance dispersal by rafting. On the other hand, the insect species of Micronesia of most orders average smaller in size than related species on nearby large land masses without showing any particular microhabitat preferences, a fact that led Gressitt (1954) to conclude that passive dispersal by

TABLE 8. Attributes that preadapt species to be good colonizers

Taxon	DISPERSAL		COLONIZATION		Large r/λ	UNKNOWN SIGNIFICANCE
	Preferred habitats superior as points of departure	High dispersal power of propagules	Preferred habitats superior as points of arrival	Large K		
Birds (Mayr, 1965a)	?	Often fly long distances as an adaptation to scattered, temporary habitats such as fresh water.	?	Often able to occupy many habitats, thus enlarging K. Tendency to eat seeds, which allows larger standing populations.	Tendency to clump (by travelling in small flocks).	
Ants (Wilson, 1961)	Tendency to occupy ecologically "marginal" (species-poor) habitats near coast.	Since marginal habitats are unstable, the species are "fugitive," with probably greater vagility.	Tendency to occupy ecologically "marginal" (species-poor) habitats near coast.	Large populations exist in marginal habitats; also typically able to occupy many habitats.	Larger average colony size.	Workers spinier, more commonly lay odor trails.
Flowering plants Dipsacaceae, Asteraceae, Rubiaceae (Ehrendorfer, 1965)	Weedy species, members of regular successions that constantly colonize newly opened habitats.	Propagules relatively widely dispersed.	Weedy species, members of regular successions that constantly colonize newly opened habitats.	?	Fast individual development, large quantities of progeny, occasional autogamy.	

wind is the major factor in insect distribution in this part of the world. The evidence for birds and ants cited in Table 8 points to a preference for unstable, scattered habitats as a preadaptation to successful colonization. As the habitats undergo succession, and competitor species begin to infiltrate the area, these fast-dispersing "fugitive species" tend to disappear. They are able to survive by the relatively quick and temporary occupancy of suitable new habitats as these first become available; in contrast, other species specialize in more persistent tenancy of relatively stable habitats. The fugitive strategy enables some species, usually a minority within a given taxon, to survive indefinitely within the kaleidoscopically changing environments of single regions. It also preadapts species to colonize neighboring regions, such as offshore islands or adjacent continents. A striking example of the relation of dispersal power to the fugitive strategy has recently been given by Maguire (1963b). In simple experimental environments, the protozoan *Colpoda* was found always to be excluded by *Paramecium*, even when it occupied the environment first. But *Colpoda* propagules also disperse farther and higher through the air than those of *Paramecium*. In a system of insular, changing environments, such as small pools, which protozoan populations are continually vacating and recolonizing, *Colpoda* could survive indefinitely.

Crowell (personal communication) has seeded various islands off the Maine coast with mice, and the fate of these is of direct relevance to the current theory. In 1962 Crowell introduced *Peromyscus maniculatus* to small islands which had only *Microtus* but which also contained habitats suitable for *Peromyscus*. He introduced 1 pair to each of three islands, 2 pairs to a second, 3 to a third, and 8 to a fourth. The 2-, 3-, and 8-pair introductions appear to have been successful (as of 1965), and one of the three single-pair ones was successful, while two left no descendants. Crowell estimates that $\lambda = 1.7$, $\mu = 1.12$, $\mu/\lambda = 0.66$, and $r = \lambda - \mu = 0.58$. $(\mu/\lambda)^x$, the probability of x pairs of individuals failing to leave ultimate descendants is $(0.66)^x$. Hence 0.66

of single pairs should fail; $(0.66)^2 = 0.44$ of two-pair introductions should fail; $(0.66)^3 = 0.29$ of three-pair introductions should fail; and only $(0.66)^8 = 0.04$ of eight-pair introductions should fail. These figures are in very good agreement with the data. Crowell's introductions have two extraordinary features that must be mentioned. First, he introduced pairs (one of each sex) simultaneously and at the same place in suitable habitat. This must be quite rare in nature. Second, his observed values of r are extraordinarily high even for established pairs. More often a species' r even in good habitat must be considerably less than 1. For a mouse with $\lambda = 5$ but r only 0.5 it would take 7 pairs to stand 50% chance of colonizing. For an island with most of the mainland species, r would probably be nearer 0.05, and it would take 70 simultaneous pairs rather than 7 to have 50% chance of colonization. This is very improbable except in the case of species that undergo periodic lemming-like irruptions or "migrations." But why are *Peromyscus* missing from all these islands when they are obviously good colonizers—once they immigrate? It may be that they are poor immigrators; perhaps, equally important, they have a small K, so that extinctions occur much more rapidly than recolonizations.

The Determinants of r

It is clearly important to a colonizing species to have a large value of r, the "intrinsic rate of increase," or more accurately, to have a large value of r/λ. It is simple to define r as $\lambda - \mu$, that is, the per capita increase in a given unit of time, and to measure it in any given interval in a real population. But how is r related over a long period of time to the schedule of births and deaths in such a real population, where different birth and death rates obtain at different ages? If $n_0(t)$ is the number of individuals of age 0 (number of newborn) at time t, and l_x and b_x are the age-specific survivorship and birth rates (i.e., l_x is the probability of a newborn surviving to at least age x, and $b_x\,dx$ is the number of offspring such a survivor has between age x and age

$x + dx$) then

(4-8) $$n_0(t) = \int_0^\infty n_0(t - \tau)l_\tau b_\tau \, d\tau,$$

for to be newborn at time t an individual must have parents who were themselves newborn at some time $t - \tau$, and these parents must have survived until age τ and then given birth. As Lotka (1925) first noticed, Equation (4-8) is satisfied by the function

(4-9) $$n_0(t) = Ke^{rt}$$

if

$$Ke^{rt} = \int_0^\infty Ke^{rt}e^{-r\tau}l_\tau b_\tau \, d\tau$$

whence

(4-10) $$1 = \int_0^\infty e^{-r\tau}l_\tau b_\tau \, d\tau.$$

From Equation (4-9) we see that $(d/dt)n_0(t) = rn_0(t)$, so that this is the same r with which we have been dealing. Equation (4-10) gives us r if only we can solve it! For specific cases it is solved by numerical integration.

Lewontin (1965), building on the work of Lotka (1925) and Cole (1954), has tabulated some values of r for various birth and death schedules, as illustrated in Figures 28 and 29. Figure 28 shows the effect of shortening the developmental time without altering the shape of the $U(x) = l_x b_x$ curve. By inspection it can be seen that a given percentage decrease of developmental time (B) is far more effective than an equivalent increase in expected total fecundity (R_0). A similar result is obtained when the $U(x)$ curve is changed in the following additional ways: B reduced without changing T or W, T reduced without changing B or W, and W reduced without changing B or T. Under each of the four extreme conditions, an appropriate alteration of the developmental time is more effective than an equivalent increase in fecundity. For example, when the intersection of the $r = 0.300$ contour with the base reference line is used as a reference point, we note that the total increase in number of offspring (R_0) required to increase r from 0.300

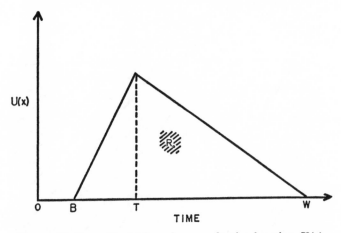

FIGURE 28. Generalized triangular reproductive function, $U(x)$ = $l_x b_x$ plotted against the age of the organism. B is the age at which first offspring are produced; T, "turnover point," the age of greatest fecundity; W, age at which last offspring are produced; R_0, expected total number of female offspring, per female parent, equal to the total area under the curve. The triangular shape of the curve approximates the actual shape determined from life history data for *Drosophila serrata* and the beetle *Calandra oryzae*. (After Lewontin, 1965.)

to 0.330 is from 780 to 1,350, nearly a doubling. The number of days needed to create the same increase, without benefit of an increase in fecundity, is given below for the four cases of change in life history:

1. Rigid translation of $U(x)$ curve   1.55-day decrease
2. Decrease age to sexual maturity
    (B only)   2.20-day decrease
3. Decrease turnover age (T only)   5.55-day decrease
4. Decrease age at last egg (W only)   21.00-day decrease

In general small changes in developmental rates of the order of 10% are roughly equivalent to large increases in fertility of the order of 100%. From this relation Lewontin predicted that relatively less genetic variance in developmental time will be found in species with a history of colonization in comparison with non-colonizing species, since developmental time will be subjected to selection

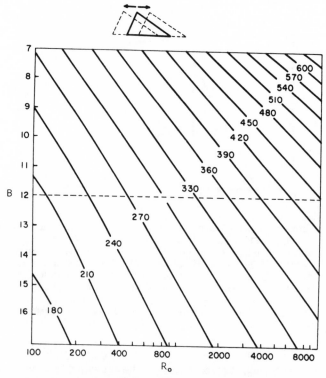

FIGURE 29. Response surface showing equal r lines for different total offspring produced (R_0) and different locations of the $U(x)$-function along the abscissa. The r values given are in parts of 1,000; thus 240 corresponds to $r = 0.240$. The dashed line corresponds to $B = 12$, $T = 23$, $W = 55$. Changes in B (the age at which first offspring are produced) shown on the ordinate are accompanied by equal changes in T and W. (From Lewontin, 1965.)

favoring a high r, whereas there should be proportionately greater variance in fecundity. Data from *Drosophila* are consistent with this prediction.

Returning now to the calculation of r, we can employ an approximate solution of Equation (4-10) which is useful in leading to further conclusions about the optimal evolutionary strategy for increasing r. As just mentioned, it has been determined empirically for at least a few species that

$U(x) = l_x b_x$ rises from zero to a maximum and then falls again. Let M be the mean age at reproduction, i.e.,

$$M = \frac{\int_0^\infty x l_x b_x \, dx}{\int_0^\infty l_x b_x \, dx},$$

so that

(4-11) $$\int_0^\infty (x - M) l_x b_x \, dx = 0.$$

Now, expand $e^{-r\tau}$ about $\tau = M$:

$$e^{-r\tau} = e^{-rM} - r(\tau - M)e^{-rM} + \frac{r^2}{2}(\tau - M)^2 e^{-rM} + \cdots$$

Substituting this into Equation (4-10) and ignoring later terms in the series, we get

$$1 = e^{-rM} \int_0^\infty l_\tau b_\tau \, d\tau - re^{-rM} \int_0^\infty (\tau - M) l_\tau b_\tau \, d\tau$$
$$+ \frac{r^2}{2} e^{-rM} \int_0^\infty (\tau - M)^2 l_\tau b_\tau \, d\tau$$

which in view of Equation (4-11) and on setting both

$$R_0 = \int_0^\infty l_\tau b_\tau \, d\tau,$$

and

$$R_0 \text{ var (age)} = \int_0^\infty (\tau - M)^2 l_\tau b_\tau \, d\tau$$

becomes

(4-12) $$1 = R_0 e^{-rM} + \frac{r^2}{2} e^{-rM} R_0 \text{ var (age)},$$

or

$$e^{rM} = R_0 + \frac{r^2 R_0}{2} \text{ var (age)},$$

where var (age) is the variance of age at reproduction. Even this equation is hard to solve for r, although it is easy graphically (Figure 30). Notice that as var (age) increases, so does r, for fixed M and R_0. If the variance in age at repro-

87

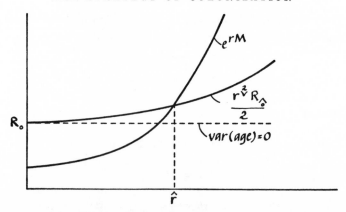

FIGURE 30. The graphical solution of r, where M, R_0, and var (age) are known and treated as constants. $\hat{r}$ is the value of r satisfying Equation (4-12).

duction is small, the solution is very simple (and well known):

$$e^{rM} = R_0,$$

whence

(4-13) $r \sim (\ln R_0)/M.$

This shows that r increases with the logarithm to the base e of the replacement rate R_0, which is defined as the total expected number of female offspring to a female newborn throughout her life; and r also increases with the reciprocal of the mean age of reproduction, M. Even this oversimplified formula lets us clear up one common misconception. It confirms that when R_0 is large, reducing M is the easiest way to increase r, as Lewontin's models indicated. However, when R_0 is near 1, a reduction in mean age at reproduction will increase r only to an insignificant extent. But, returning to the problem of effective colonization, we desire a species that can most readily achieve large values of r/λ. It is clear that the good colonizer has a large r, which is achieved by a low mortality rate rather than by a high birth rate. In fact, it is the ratio λ/μ, rather than the difference $r = \lambda - \mu$, which we would expect to be maximized in a good colonizer.

The Reproductive Value of Propagules

A further concept from demography—reproductive value (Fisher, 1930)—tells us at which age a propagule is most likely to establish a successful colony. In biogeographic terms, the reproductive value v_x of an x-year-old may be defined as the expected size of a colony (at some remote future time) founded by a propagule of x-year-olds. We can then divide this number by the size of a simultaneous colony founded by a propagule of newborn individuals to make the definition independent of the exact time interval before the count is made. The colony founded by newborns will be smaller for two reasons: first some will have died before reproducing; second the colony founded by the older propagules will have had a head start because their exponential increase will have started sooner. Motivated by these considerations, Fisher (1930) defined the reproductive value, v_x, of an x-year-old in an expanding population so that its ratio to v_0 satisfies

$$\frac{v_x}{v_0} = \int_x^\infty e^{-r(y-x)} \frac{l_y}{l_x} b_y \, dy = \frac{e^{rx}}{l_x} \int_x^\infty e^{-ry} l_y b_y \, dy.$$

Here $(l_y/l_x)b_y dy$ can be looked on as the expected number of newborn to an individual now of age x in the time interval y to $y + dy$, because l_y/l_x is the probability of an x-year-old reaching age y. The term $e^{-r(y-x)}$ is a weighting factor measuring the advantage of early reproduction. Immediate reproduction $(y = x)$ is weighted $e^{r(y-x)}$ times as much as reproduction at a time $y > x$, since an individual alive now has left $e^{r(y-x)}$ descendants after time $y - x$. We integrate from x to ∞ to count only the remaining births. The motivation is rather heuristic, and the real justification for the term comes from the following mathematical considerations. (The reader already convinced that a propagule is more likely to succeed when its individuals are of the age of maximum reproductive value may wish to skip the formal explanation presented in the next paragraph.)

Fisher proved that, no matter what the age distribution of individuals present in the population, the reproductive value of x-year-olds plus their line of descent always grows according to e^{rt}. This is quite remarkable, for the number of individuals only asymptotically reaches a smooth exponential growth and initially grows very irregularly, depending on how many of the individuals are of reproducing age. In spite of this, the reproductive value always grows smoothly.[3] Fisher proved this for continuously growing populations; both the argument and the computations are considerably more transparent in the discrete case, so we follow that approach. We must, therefore find v_0, v_1, v_2, v_3, . . . , which are the reproductive values of newborns, of 1-year-olds, of 2-year-olds, etc. (Of course these could be 1-day-olds, 2-day-olds, etc. instead, if the time unit is one day instead of one year). Instead of beginning with a definition of reproductive value, we begin with the property we wish it to have. An x-year-old this year will, by next year, have survived to become an $x + 1$ year-old with probability l_{x+1}/l_x. For definiteness we assume the population counts are made just following reproduction; in other words our x-year-old does not reproduce again until just preceding age $x + 1$, and it must survive in order to do this. Hence our x-year-old, with reproductive value v_x (although we do not know what v_x is, yet) will contribute in two ways to the reproductive value of the population at the next time interval. It becomes an individual of value v_{x+1} with probability l_{x+1}/l_x and, with the same probability, produces b_{x+1} newborn[4] of value v_0. But we want this new reproductive value $(l_{x+1}/l_x)(v_{x+1} + b_{x+1}v_0)$ to be e^r times the present value, v_x. In symbols, we want v_x to be defined so that

$$(l_{x+1}/l_x)(v_{x+1} + b_{x+1}v_0) = e^r v_x.$$

[3] Some of the mystery of the term disappears when it is viewed in the formalism of linear algebra where r becomes an eigenvalue and v_x defines an eigenvector.

[4] b_x is slightly different here, being the expected number of newborn during the unit time interval, rather than a continuously varying number.

Since v_x is relative to v_0, we are at liberty to set $v_0 = 1$. Then we can rearrange the equation to get $v_{x+1} = (l_x/l_{x+1})e^r v_x - b_{x+1}$, which is a recursive definition for v_{x+1} in terms of v_x. The first few values are

$$v_0 = 1$$

$$v_1 = \frac{l_0}{l_1} e^r v_0 - b_1 = \frac{e^r}{l_1} - b_1 \text{ (because } l_0 = 1, v_0 = 1)$$

$$v_2 = \frac{l_1}{l_2} e^r v_1 - b_2 = \frac{e^{2r}}{l_2} - \frac{l_1 b_1 e^r}{l_2} - b_2$$

$$v_3 = \frac{l_2}{l_3} e^r v_2 - b_3 = \frac{e^{3r}}{l_3} - \frac{l_1 b_1 e^{2r}}{l_3} - \frac{l_2 b_2}{l_3} e^r - b_3$$

.

.

.

$$v_x = \frac{e^{xr}}{l_x} - \frac{l_1 b_1 e^{(x-1)r}}{l_x} - \frac{l_2 b_2}{l_x} e^{(x-2)r} - \cdots - b_x.$$

These values, defined so as to make the reproductive value of the population grow according to e^{rt}, give the easiest means of calculating the v_x, recursively. To obtain Fisher's formula, we notice that since r is defined so that

$$e^{xr} = \sum_{i=1}^{\infty} l_i b_i e^{-(i-x)r} = \sum_{i=1}^{x-1} l_i b_i e^{-(i-x)r} + \sum_{i=x}^{\infty} l_i b_i e^{-(i-x)r},$$

we obtain

$$l_1 b_1 e^{(x-1)r} + \cdots + l_{x-1} b_{x-1} e^r = e^{xr} - \sum_{i=x}^{\infty} l_i b_i e^{-(i-x)r}.$$

Substituting this into the formula for v_x,

$$v_x = \frac{e^{xr}}{l_x} - \frac{1}{l_y}\left[e^{xr} - \sum_{i=x}^{\infty} l_i b_i e^{-(i-x)r} \right] = \sum_{i=x}^{\infty} \frac{l_i b_i e^{-(i-x)r}}{l_x}$$

which is Fisher's formula.

This completes the proof that v_x, defined according to Fisher, grows at the rate e^{rt} and can be computed from simple recursive formulae.

Turning back to the strategy of colonization, a propagule

of x-year-olds has a better chance of surviving than an equal sized propagule of y-year-olds if $v_x > v_y$, regardless of whether x is greater or less than y. For this reason we would expect evolution to favor dispersal at the age of maximum reproductive value, which usually approximates the age at which reproduction commences. The nuptial flights of queen ants would be an obvious case of this strategy. For birds, by contrast, l_x/l_{x+1} is independent of x and so is b_x, as Lack (1954) has shown. In other words, an x-year-old is indistinguishable from an $x + 1$-year-old. Hence for birds v_x is about constant and any age propagule is as good as any other. The exact advantage accruing to the colonist with the maximum reproductive value is not yet clear and can stand as a challenge to mathematicians.

SUMMARY

In order to predict the probability that a propagule of a given species will establish a successful colony, it is necessary to be able to predict the longevity of populations from a knowledge of the basic demographic parameters. In an attempt to achieve this, a model has been constructed relating per capita birth rate (λ), per capita death rate (μ), and equilibrial population size attainable on the island (K) to the probability of survival of propagules and the average duration of populations. Equations were derived under certain conditions of density-dependent population growth, and these were then employed to make the first qualitative generalizations about the fate of colonizing populations. The following conclusions were drawn:

(1) The chance of an individual propagule leaving descendants which eventually grow to the maximum population size (K) is about r/λ, where r is $\lambda - \mu$, the intrinsic rate of population increase.

(2) The average time a propagule and its descendants take to go extinct can be calculated from a specified λ, μ, and K. Curves relating these parameters are given in Figures 27A, 27B, and 27C. From the figures it can be seen that there is a "take-off" value of K for every λ and μ above

which the average survival time increases rapidly and soon becomes enormously large.

(3) The average survival time of a population already at K is about λ/r times the average survival time of a propagule and its descendants.

Since small changes in λ and μ can have such a great effect on probability of colonizing success, the determinants of these parameters have been examined in greater depth. In general, probability of colonizing success increases very rapidly with an increase in λ/μ but not necessarily so with merely an increase in the absolute difference $r = \lambda - \mu$. Moreover, speeding the developmental rate and increasing the span over which reproduction can occur are usually more effective than increasing fecundity. At the level of the individual organism, the most effective propagules are those at the age of maximum reproductive value; this age and the reproductive value can be defined precisely from demographic data.

From these *a priori* mathematical considerations, a biological portrait of the superior colonist is drawn and matched against empirical descriptions of superior colonizing species made by previous biogeographers.

CHAPTER FIVE

Invasibility and the Variable Niche

In the last chapter we dealt with the difficulties that face a colonizing propagule even when conditions favor the population's increase. In our models the colonist was allowed to establish a beachhead, however briefly, and subsequent extinction was treated as a random variable. Now we turn to situations that prevent the colonist from successfully invading no matter how many times it tries. A community that cannot be invaded by a given species is "closed" to that species. Another, familiar way of putting it is to say that the species' "niche" is already filled by other species. Yet this particular description is ambiguous and often inaccurate, since a species' ecological role is plastic, and it can be closed out of a community by species that bear little morphological or behavioral resemblance to it. Furthermore, it may successfully invade even if its niche is full, providing it is a superior competitor.

This chapter has two aims. First it will be established that a community can be closed to a species in ways that are often both difficult to predict and surprising in their consequences. Second, attention will be paid to another aspect of the same problem: the kinds of pressures exerted by competitors and the relationship of ecological plasticity to colonizing success.

The Closed Community

Volterra (1926) was the first to point out that there can be no more predator species than prey species in a given habitat, provided the predator species are resource limited. ("Prey" here is interpreted to include the plant food of herbivores.) A generalized version of Volterra's observations developed in models by MacArthur and Levins (1964, and in press) is the basis of the present discussion. Suppose predator species populations are designated P_1, P_2, . . . , P_n and resource species populations R_1, R_2, . . . , R_m. The predator species will increase when their particular resources exceed some threshold density, F_i:

$$(5\text{-}1) \qquad \frac{\mathrm{d}P_i}{\mathrm{d}t} = P_i f_i \left[\sum_j a_{ij} R_j - F_i \right],$$

where f_i is some monotonic function such that $f_i(0) = 0$ and a_{ij} is a constant. Since the P_i are resource limited there are no P terms in the bracket. When f_i is the identity function (i.e., the symbol f_i can be removed), Volterra's equations are obtained. Even more generally, $\mathrm{d}P_i/\mathrm{d}t$ can be written as follows:

$$(5\text{-}2) \qquad \mathrm{d}P_i/\mathrm{d}t = P_i f_i[g_i(R_1, R_2, \ldots, R_m) - F_i],$$

where g_i is another function that increases with each R_i. Equation (5-2) clearly includes Equation (5-1) when g_i takes on the values of the sum in Equation (5-1). In either case, there is a surface $g_i(R_1, R_2, \ldots, R_m) = F_i$, in the space whose coordinates are R_1, R_2, . . . , R_m, which marks the inner edge of the zone in which P_i increases. Each species has such a surface, and since m such surfaces intersect at a point, no more than m predators can simultaneously persist on the m resources. This theorem is illustrated for the cases $m = 1$ and $m = 2$ in Figure 31. Of course more species can be added if the competitors are also able to divide up parts of the habitat, although even this increase must approach an upper limit. The theorem we have just proved applies to resource-controlled species that occupy the same habitat.

To extend the result let us now introduce the concept of the grain size of resources. If two resources come in a mixture such that the predator encounters them in the proportion in which they occur, they are called "fine-grained." If the predator is able to search in such a way as to come upon[1]

---

[1] This is not to say the species does not select from among the resources it encounters. This is a slight modification of the original definition of MacArthur and Levins (1964).

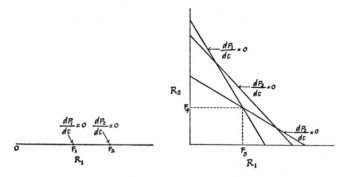

FIGURE 31. Diagrams illustrating the theorem that m resource species can support no more than m predator (or herbivore) species provided the latter are resource-limited. The left-hand diagram gives the case of $m = 1$. R_1 is the density of resource of type 1. F_1 is that quantity of R_1 below which predator species 1 cannot increase, and F_2 is the similar threshold quantity for predator species 2. Since species 1, being resource-limited, will crop the resource species down to F_1, species 2 will decrease (i.e., $dP_2/dt < 0$) and ultimately disappear. The right-hand diagram gives the case of $m = 2$. Now the threshold values are lines relating R_1 and R_2. Species 1 and 2 possess the innermost intersect (F_3, F_4), which falls inside the threshold line of species 3; hence $dP_3/dt < 0$, and species 3 cannot invade a community containing species 1 and 2.

one resource in more than its natural proportion, these resources are "coarse-grained"; i.e., this predator can either select the grains of environment which contain one kind of resource or else spend its whole life on a grain of one resource, even if that grain is selected randomly. Notice that "fine" and "coarse" are relative to the size of the predator: different species of herbs that insects might select in a coarse-grained fashion would be encountered indiscriminately— hence in a fine-grained fashion—by some large grazer. In terms of grain the relevant points are these: First, a single fine-grained predator may occasionally out-compete two coarse-grained specialists, if these waste a considerable amount of time by travelling across unsearched areas and "selecting" the particles of proper resource. Second, only one predator can persist on two fine-grained resources if these are indiscriminately consumed and simultaneously reduced in abundance, since the predator who reduces them to the lowest level wins. Hence, under certain conditions, two resources may support only a single species of predator.

Summarizing, there can be no more predators in a given habitat than the kinds of resource on which they depend, and there may be fewer if the resources come in fine-grained mixtures. Thus there is an upper limit to the number of resource-limited species the habitat will hold. The number of species in the whole community cannot exceed the number per habitat times the number of recognized habitats.

There are two other ways in which a community can be closed to an invader. First, on an island there is a limited number of individuals of all species combined. For instance, Crowell (1961) found that three mainland bird species native on Bermuda maintain as great a combined population as those three plus many others in the comparable habitats on the mainland. There is much other anecdotal evidence of this phenomenon, which is the most convincing evidence of competition. Conversely, as new species are added to a fauna, each must become rarer, unless, as in the case of the more recently introduced starling and kiskadee on Bermuda, the species are able to tap a previously unused food supply. When the number of species grows too large, each will therefore have a small abundance and will rather quickly go extinct by the means described in the first section of Chapter 4. Schoener (1965) has exploited this view in his interpretation of character differences among insular bird species, and it is also a postulate of our equilibrium model.

Finally, predators can switch to common foods, and disease epidemics can break out when the target species are sufficiently common. Hence, in the presence of predators or disease, no prey can safely be too abundant, while in the absence of predators or disease one or two resource species may become dominant, excluding others. Paine (1966, and personal communication) has conducted an experiment which demonstrates this principle in impressive fashion. He studied a marine rocky shore community at Mukkaw Bay, Washington, which, in the presence of the starfish predator, *Pisaster*, had the food "sub-web" shown in Figure

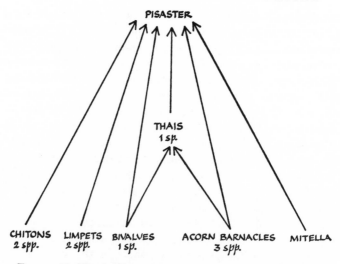

FIGURE 32. The feeding relationships of the food sub-web at Mukkaw Bay, Washington, dominated by the starfish *Pisaster ochraceus*. Removal of *Pisaster* resulted in increase of certain prey species to the detriment of others, and a reduction in species density. (Modified from Paine, 1966.)

32. The sub-web contains all species from which energy finds its way into *Pisaster*. Paine removed the starfish from a piece of shore 8 m × 10 m in area and kept the area free of them afterward. On removal of the predatory starfish, space became short, and, successively, barnacles of the genus *Balanus*, then the bivalve *Mytilus californianus* crowded other species out. The actual number of species was reduced from 15 to 8 by removing the predator. How can we interpret Paine's results? First, when space limited the lowest trophic level shown, only as many species could persist as used patches of the space differently or which could persistently reimmigrate. This accounts for the few species after removal of starfish. But why were there so many before, when the starfish kept populations sparse? Surely one was better at avoiding predation, and this one should out-compete the others. Most likely the answer is contained in either continued immigration of the low trophic levels into the space kept clear by the predator or else the

existence of a greater number of operating limiting factors in the presence of the predator. Slobodkin (1961) noted a theoretical explanation for effects of this type.

Each of the three conditions for reduction of diversity— competitors too similar, species too rare, predators too few (or too many)—can prevent invaders from colonizing successfully. In only a few cases would their influence be apparent to the biogeographer without the benefit of an extended ecological study. There certainly exist some striking cases of distributional gaps that call for such deeper studies. For example, the bananaquit (*Coereba flaveola*) is among the most abundant bird species on almost all of the larger islands of the West Indies. On Puerto Rico, in particular, it is found in every habitat but grassland and is the commonest species in most of these habitats. It is certainly vastly commoner than any other Puerto Rican bird, as shown by the censuses from MacArthur, Recher, and Cody (1966) reproduced here in Table 9. A similar pattern is typical of other islands from the Lesser Antilles to the Bahamas and the Cayman Islands and Jamaica, with two notable exceptions—there are no bananaquits on Cuba or the nearby Isle of Pines! It is inconceivable that no bananaquits have ever immigrated onto Cuba, since they have reached all other islands both north, south, and east of Cuba, and Cuba is much the biggest target. Furthermore, there are no obvious counterparts to the bananaquit on Cuba, although the island does have the red-legged honeycreeper (*Cyanerpes cyaneus*) which, like the bananaquit, feeds from flowers. However, not only is this honeycreeper mainly confined to the mountains in Cuba, but it is altogether missing from the Isle of Pines, so by itself it cannot restrict the bananaquit very much. Furthermore, it occurs with bananaquits on islands smaller than Cuba, as we shall presently see. We can only conclude that Cuba's present bird fauna is sufficiently complete that r and K for an immigrant bananaquit are small, and that the frequency of propagules of adequate size is vanishingly small. It is hard to say what species really "should" be on Cuba, since it is so remote from its various sources of colonization.

TABLE 9. Puerto Rico breeding bird censuses. (From MacArthur, Recher, and Cody, 1966.)

Species	El Verde forest	Mount Britton forest	Luquillo grassland	Lajas Valley savannah	Guanica desert	Maricao forest	Estacion Rubias cultivated
Columba squamosa	2(.5)	1(.5)					
Zenaida aurita				1(1)	1(1)		
Columbigallina passerina				6(6)	5(5)		
Geotrygon montana	2(1)						
Amazona vittata	+						
Saurothera vieilloti	1(0)	+					.5
Chlorostilbon maugaeus	2(.5)	1.5(1)				1	2
Anthracothorax dominicus					2(2)		
A. viridis							2
Todus mexicanus	6.5(1)	2.5(1.5)			2(1)	3	1
Melanerpes portoricensis	.5(0)	+				1	
Tyrannus dominicensis				6(5)	3(2)		
Myiarchus stolidus				2(1)	1(1)		
Contopus latirostris							+
Mimus polyglottos				3(3)	1(1)		
Margarops fuscatus	6(1)	2(2)					
Mimocichla plumbea						1	
Vireo latimeri							
V. altiloquus	17(4.5)			2(2)	2(1)	5	
Dendroica adelaidae				5(4)	1(1)		.5
Coereba flaveola	81(13.5)	26.5(17.5)		2(1)	8(7)	5	
Tanagra musica					5(2)	1	10

TABLE 9. (Continued)

Species	El Verde forest	Mount Britton forest	Luquillo grassland	Lajas Valley savannah	Guanica desert	Maricao forest	Estacion Rubias cultivated
Spindalis zena	6(0)	1(0)			2(1)		3
Nesospingus speculiferus	2.5(.5)	1(0)			.5(.5)	3	3
Quiscalus niger				1(0)			
Icterus dominicensis				.5(0)			
Loxigilla portoricensis	3(.5)	2.5(1.5)	31(20)		1(1)	1	5
Tiaris olivacea			8(5)				+
T. bicolor				2(2)	.5(.5)		1
Total individuals	(23) 129.5+	(25) 38+	(25) 40	(25) 30.5	(27) 35	21	28+
Total species	12+	8+	2	11	15	9	11+

Notations: + indicates the species was present but spent an insignificant portion of its time in the census plot.
Numbers outside parentheses indicate total numbers of territories, including fractions of territories in some cases.
Numbers in parentheses indicate numbers of territories in a subplot large enough to contain about 25 pairs. This figure allows a comparison between habitats which differed in absolute bird densities.

101

TABLE 10. Panama breeding bird census. Conventions are the same as in Table 9. (From MacArthur, Recher, and Cody, 1966.)

Species	Mature forest	Second growth	High savannah	Low savannah	Mangrove	High grass	Low grass
Buteo magnirostris				.05(.02)			
Spizaetus tyrannus	+(+)						
Ortalis garrula		2(0)					
Colinus cristatus			1(1)	1(1)			
Columba cayennensis		1(0)					
Zenaida macroura			1(0)				
Columbagallina talpacoti		1(0)					
Claravis pretiosa		1(0)					
Leptotila verreauxi		8(2)					
L. cassinnii	1(1)						
Aratinga pertinax			4(4)	3(1.5)			
Tapera naevia		5(2)	1(1)				
Piaya cayana		1(0)					
Otus choliba		.1(0)					
Phaethomis superciliosus	1(1)						
Chlorostilbon canivetii		1(0)	2(1)	2(0)			
Amazilia amabilis			1(1)				
A. tzacatl		2(0)					
hummingbird (sp ?)	1(1)						
Trogon massena	.5(.5)						
Ramphastos sulfuratus	1(.3)						
R. swainsonii	.5(.5)						
Melanerpes pucherani	2(0)						
M. rubricapillus			1(0)		2		
Dryocopus lineatus	1(0)						
Dendrocincla fuliginosa	1(0)				1		
Glyphorhynchus spirurus	2(1)						
Xiphorhynchus guttatus	1(1)						
X. lachrymosus	.5(.5)						
Xenops minutus	4(3)						
Thamnophilus punctatus	5(1)	4(2)					
T. doliatus		2(0)					
Dysithamnus puncticeps	2(1)						
Myrmotherula axillaris	6(4)						
Microrhopias quixensis	3(0)						
Cercomacra tyrannina	1(0)						
Myrmeciza longipes		1(0)					
Hylophylax naevioides	1(0)						
Attila spadiceus	1(0)						
Rhytipterna holerythra	1(0)						
Tityra semifasciata	.5(0)				1		
Querula purpurata	1(0)						
Pipra mentalis	4(2)						
Manacus vitellinus		3(1)					
Chiroxiphia lanceolata		2(0)					
Muscivora tyrannus				7(6)			
Tyrannus melancholicus		1(0)	2(1)	2(2)			
Legatus leucophaius	1(0)						
Megarynchus pitangua		1(0)					
Myiozetetes similis		2(1)	3(2)	1(0)	1		
Myiarchus ferox	3(0)	1(0)			4		
M. tuberculifer	1(0)						
Terenotriccus erythrurus	1(1)						
Cnipodectes subbrunneus		1(0)					
Tolmomyias assimilis	1(0)						
Todirostrum cinereum					2		
Phylloscartes flavovirens		1(0)					
Capsiempis flaveola		2(0)					

TABLE 10. (*Continued*)

Species	Mature forest	Second growth	High savannah	Low savannah	Mangrove	High grass	Low grass
Elaenia flavogaster		3(1)	2(2)				
E. chiriquensis		2(1)	?(?)				
Sublegatus arenarum		4(3)			3		
Tyranniscus vilissimus	1(0)						
tyrannulet (sp ?)	2(0)						
Piprimorpha oleaginea	2(0)						
Troglodytes musculus				1(0)			
Thryothorus leucotis	1(1)						
wren (sp ?)	1(0)						
T. rutilus		1(0)					
Cyphorhinus phaeocephalus	1(0)						
Turdus grayi		4(1)	3(2)	1(0)			
Smaragdolanius pulchellus	1(0)						
Hylophilus decurtatus	6(2)						
Vireo flavoviridis		8(3)			9		
Dacnis cayana	1(0)						
Cyanerpes cyaneus	1(0)	3(1)					
Basileuterus delattrii		1(0)					
Amblycercus holosericeus		2(0)					
Icterus chrysater		1(0)					
Sturnella magna				1(1)		3	
Leistes militaris						6	
Molothrus aeneus				1(1)			
Thraupis episcopus		1(1)	3(3)				
Rhamphocelus dimidiatus			2(1)				
Eucometis penicillata	1(0)						
Saltator maximus		2(0)					
S. albicollis		4(2)	1(0)				
Sporophila aurita		1(0)	1(1)				
S. minuta				6(4)			
Volatinia jacarina		4(1)	8(6)	26(11)		12	
Tiaris olivacea							1
Arremonops conirostris		2(2)					
Zonotrichia capensis							8
Emberizoides herbicola							2
Total individuals	(21.8) 66.5	(25) 88.1	(25) 34	(28.52) 55.05	23	21	11
Total species	40	39	15+	14	8	3	3

Other floristically diverse islands nearer their sources of colonization are perhaps more illuminating. For instance the island of Coiba off the Pacific Coast of Panama presents several unexpected lacunae in its bird fauna (Wetmore, 1957). This island is barely 15 miles from the nearest point of the Panamanian mainland, is large (21 × 15 miles), and rises to a height of 1,400 feet. It is well watered and is covered with exceptionally tall, undisturbed forest. From all appearances such an island might be expected to support most of the forest species of the mainland. Actually, it has only 94 species of resident birds, which is only a small fraction of the mainland forms (although both bananaquits and red-legged honeycreeper coexist in abundance!). In spite of the short distance to the mainland, some evolution has occurred. Wetmore could distinguish as subspecifically distinct no less than 20 of the 94 species. What is of interest here, however, is the missing families of birds. Coiba has no tinamous (2), curassows and guans (3), trogons (5), motmots (3), jacamars (0), puffbirds (6), toucans (3), or wood-hewers (6). The numbers in parentheses are the numbers of species on smaller, less diverse, Barro Colorado in the Panama Canal (Eisenmann, 1952). Some other families are almost missing: there is only one ovenbird (3), one antbird (17), one manakin (4), and one wren (7). This impoverishment cannot be accounted for in terms of impoverished island habitat. Rather, provided equilibrium has been reached, r and K must be sufficiently low that extinctions happen quite rapidly. Hence it appears that the second mechanism for preventing invasion (species too rare) might be the primary one on islands.

Short-term Changes in Food and Habitat

A species that successfully colonizes an island has entered a new environment. Not only are physical conditions usually different in some way, but there will always be a different combination of species, or relative abundances in the species, or both, in the invaded insular environment. As a consequence the colonizing species will probably experience one or the other of two kinds of immediate ecological change. Either it will shift its preference, or it will undergo some form of expansion or contraction. These changes might be purely phenotypic at first, reflecting the species' behavioral or morphological plasticity, and be translated into genetic differences later by natural selection, perhaps involving genetic assimilation. We will confine discussion for the moment to potential immediate phenotypic change and take up the implication of evolutionary reinforcement in a later section. Reports of ecological change, involving unknown proportions of phenotypic and genetic components, are commonplace in the literature (e.g., Lack, 1942; Wilson, 1959; Mayr, 1963). The changes are sometimes referred to as ecological displacement (shift or contraction) and ecological release (expansion).

Ecological Expansion and Contraction

Compressibility of an elementary kind is illustrated in the comparison between the Puerto Rico and Panama bird censuses given in Tables 9 and 10. Clearly the Puerto Rican birds are found in many habitats, while the Panamanian ones are much more restricted. A diversity of similar effects is shown by tropical ant faunas (Wilson and Taylor, 1967; Wilson and Hunt, 1967). The ants of the Wallis Islands, which are located between Fiji and Samoa, exemplify the complex changes that accompany reduction in species diversity. The native fauna of Uvéa, the principal island of the group, is a small subset of the native Fijian fauna. Its species occur in different abundances from the conspecific populations on Fiji. Surrounding Uvéa are 20 islets, which in turn are populated by small subsets of the Uvéa species. Despite the fact that the islets are strung close together on a coral reef and are similar in appearance, their faunulae show marked differences in the composition, relative abundances, and even nest sites of the resident species.

Entry into a smaller fauna is often accompanied by ecological release. On New Guinea some of the most widespread of the Indo-Australian ant species, notably *Rhytidoponera araneoides, Odontomachus simillimus, Pheidole oceanica, P. sexspinosa, P. umbonata, Iridomyrmex cordatus,* and *Oecophylla smaragdina,* are mostly or entirely limited to species-poor "marginal" habitats, such as grassland and gallery forest. But in the Solomon Islands, which has a smaller native fauna, these same species also penetrate the rain forests, where they are among the most abundant species. In the New Hebrides, which has a truly impoverished ant fauna, the species of *Odontomachus* and *Pheidole* just listed almost wholly dominate the rain forests as well as the marginal habitats. Ecological release in the opposite direction, from central to marginal habitats, has also

105

occurred. In Queensland and New Guinea, *Turneria* is a genus of rare species mostly confined to rain forests. It is also the only genus of the subfamily Dolichoderinae to have reached the northern New Hebrides. On Espiritu Santo, one of the latter islands, two species of *Turneria* are among the most abundant arboreal insects in both marginal habitats and virgin rain forest.

The degree of compression or release in new environments varies among species and is difficult to predict in advance. A case in point is the marked difference in behavior between two of the thirteen ant species that have succeeded in colonizing the Dry Tortugas, the outermost of the Florida Keys. In the presence of such a sparse fauna, *Paratrechina longicornis* has undergone extreme expansion. In most other parts of its range it nests primarily under and in sheltering objects on the ground in open environments. On the Dry Tortugas it is an overwhelmingly abundant ant and has taken over nest sites that are normally occupied by other species in the rest of southern Florida: tree-boles, usually occupied by species of *Camponotus* and *Crematogaster*, which are absent from the Dry Tortugas; and open soil, normally occupied by the crater nests of *Conomyrma* and *Iridomyrmex*, which genera are also absent from the Dry Tortugas. In striking contrast is the behavior of *Pseudomyrmex elongatus*. This ant is one of ten species that commonly nest in hollow twigs of red mangrove in southern Florida. It tends to occupy the thinnest twigs near the top of the canopy and is only moderately abundant. *P. elongatus* is also the only member of the arboreal assemblage that has colonized the Dry Tortugas, where it has a red mangrove swamp on Bush Key virtually all to itself. Yet it is still limited primarily to thinner twigs in the canopy and, unlike *Paratrechina longicornis*, has not perceptibly increased in abundance.

Cameron (1958) has described an interesting case of expansion followed by compression in the arctic hare (*Lepus arcticus*), which was initially the only hare on Newfoundland, where it expanded to occupy both its mainland tundra habitat and also forest land. When the varying hare (*Lepus americanus*) was introduced, the habitat of the arctic hare contracted back to the tundra.

Once compressibility of any degree has been demonstrated, the next general question to ask is whether diet or habitat or both should contract when competitors are present. The converse question is whether diet or habitat or both should expand where interspecific competition is relaxed. The theoretical answer is quite interesting: when competition is increased, the variety of occupied habitats (or more correctly, the space searched) should shrink, or at least be altered, but the range of foods within the occupied habitats should not. MacArthur and Pianka (1966) give a formal proof of this hypothesis, but the following verbal argument is adequate. (1) On hunting for food within a fine-grained patch of habitat an efficient species should clearly accept or reject each item encountered on its own merits, which are nearly independent of how rare other such items of the same species are. That is, the choice is made item by item as the items of food are found; and the finding within a searched area is not controlled by preference of one food or another. Hence if a new competitor enters the patch and reduces the density of a particular food species, it will not greatly affect whether a previous species includes that food in its diet, although it may reduce the *proportion* of the food in the diet. (2) However, each habitat may be thought of as a mosaic of fine-grained patches. Which of these patches a species hunts depends very much on competition, because this choice is made ahead of time, and, by avoiding a patch that is heavily foraged by a competitor, a species can increase its harvest of food. Actually, competition may not only restrict the types of patches in which the species feeds; it may also cause new, previously unsuitable patch types to be included in the itinerary. This is quite consistent with the hypothesis and has in fact been demonstrated in ant thrushes by Willis (1966). In sum, a species can, without marked waste of time, forage only in the more profitable kinds of patches. But within each patch it cannot avoid coming upon the full spectrum of food, since the patches are fine-grained, and thus be forced to decide independently upon each item. Consequently, on being freed from competition on an island, a species can be expected to alter and usually to enlarge its habitat, but not its range of diet—at least initially—although the variance of items in the diet may be enlarged. Looked at another way, as a new island is occupied by progressively more and more species, both the preferred food and habitat should be initially as varied as the species' morphology allows, but the habitat primarily should change with the entry of new species. There are limits to this process. As the species rejects more and more kinds of patch, in favor of the most productive kind, it eventually begins to lose an appreciable amount of time in travelling. Also, as it restricts the patches in which it feeds, it may automatically eliminate some kinds of food and thus alter its diet. At this point, further habitat restriction is unlikely. The relationships can be simply visualized in terms of species packing and unpacking, as shown in Figure 33.

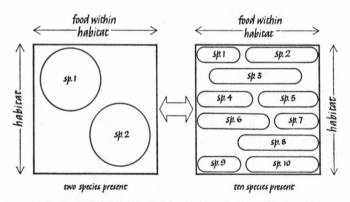

FIGURE 33. The compression hypothesis. As more species invade and are packed in, the occupied habitat shrinks (although some marginal patches may be added), but not the range of acceptable food items within the occupied habitat. The actual diet, reflecting both acceptability and abundance of items, may become more concentrated, but the range of items should not greatly change. Conversely, as species invade a species-poor island from a species-packed source, only the occupied habitat expands. The hypothesis applies only to short-term, non-evolutionary changes.

108

An empirical documentation of habitat patch selection has been provided by J. Bruce Falls and his student John D. Featherstone (personal communication), who applied insecticide, and thereby reduced the insect density, in a checkerboard pattern within six ovenbird territories. The bird foraging was dramatically reduced in the squares of reduced food density.

Part of the packing effect, namely habitat expansion, has been amply demonstrated by field studies, several of which have already been cited. Recently MacArthur, Recher, and Cody (1966) have undertaken a closer analysis of habitat expansion in the loosely packed Puerto Rican bird fauna. These authors found that not only are the species expanded horizontally, but they also appear to be less restricted to certain feeding heights above the ground. To make these observations precise, a measure was employed "FH diff" (foliage height difference) of difference between habitats and a measure "BS diff" (bird species difference) of difference between the bird communities of two habitats. The *bird species diversity*, H_1 of area 1 is defined by $H_1 = \sum_i p_i \ln p_i$ where p_i is the proportion of all individuals which are of species i, so that $\sum_i p_i = 1$. Similarly H_2 is defined for habitat 2: $H_2 = \sum_i p_i' \ln p_i'$. We also form an H_T for the total combined census, as follows:

$$H_T = \sum_i \left(\frac{p_i + p_i'}{2}\right) \ln \left(\frac{p_i + p_i'}{2}\right),$$

where i runs through all species found in either habitat and therefore p_i' or p_i may be zero if a species is missing from one habitat. We now define BS diff $= H_T - (H_1 + H_2)/2$. BS diff varies from 0 (when the same species are found in the same relative abundance in both areas) to $0.693 = \ln 2$ (when there are no species in common to the two areas). Similarly FH diff (foliage height difference) is measured by $H_T - (H_1 + H_2)/2$ where this time H_1 (which we call

109

foliage height diversity) $= \sum_i g_i \ln g_i$, and g_i is the proportion of the total foliage lying in the ith horizontal layer. The layers used are so determined as to make the bird species diversity, H, be determined as fully as possible by the foliage height diversity. It turned out that two layers (0–2′ above ground, >2′) were optimal in Puerto Rico, three layers (0–2′, 2–25′, >25′) in temperate United States, and four layers (0–2′, 2–10′, 10–50′, >50′) in the Canal Zone and adjacent Panama (Figures 34 and 35). Hence fewer layers are used where there are fewer species. These layers presumably correspond to foliage configurations—herbaceous, bush, canopy, etc.—and should be adjusted in other localities as the heights of these configurations change. For instance if grass goes to 3 ft, the bottom layer should be 0–3 ft. Returning to the habitat differences, we can now plot BS diff against FH diff based on two layers in Puerto Rico, three in the United States, and four in Panama (Figure 36), which shows that the same difference between habitats causes less change per layer in bird species in Puerto Rico than in either Panama or temperate United States. This documents the increased range of acceptable habitat on islands. On the other hand, it does not necessarily follow from the differences in the curves that *individual species* have expanded their habitats, for the Puerto Rican birds could be a sample of wide-ranging mainland species. Probably some such biased sampling together with some expansion occurs. Many of the wide-ranging species are endemic, so evolution may have played a large role in producing the expansion. [Future studies of this sort may use the expressions $1/\Sigma p_i^2$ and $(2\Sigma p_i p_i')/(\Sigma p_i^2 + \Sigma p_i'^2)$ for diversity and overlap, respectively. See for example Horn (1966) and Levins (1966).]

The other part of the compression theorem—the lack of reduction in range of food—is harder to prove. Notice, however, that in Figures 34 and 35 we showed the effect of additional species as restricting habitat but not food. Hence, in a single patch of habitat there should be nearly as many species on an island as on the mainland. To see this draw a

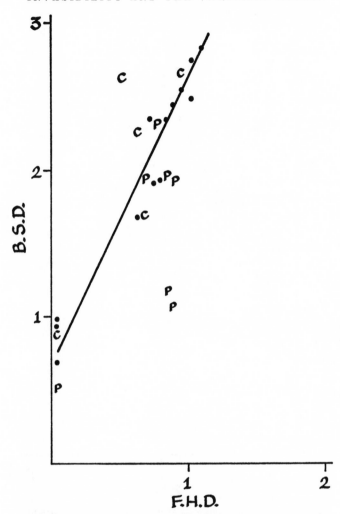

FIGURE 34. Bird species diversity (B.S.D.) is plotted against foliage height diversity (F.H.D.) for censuses from Puerto Rico (points marked with P), areas in or near the Canal Zone (points marked with C), and temperate United States (solid points and regression line). Foliage height diversities were all calculated from the *three layers* 0–2′, 2′–25′, >25′. Notice the large scatter of P and C points, and then compare this result with that in Figure 35. (After MacArthur, Recher, and Cody, 1966.)

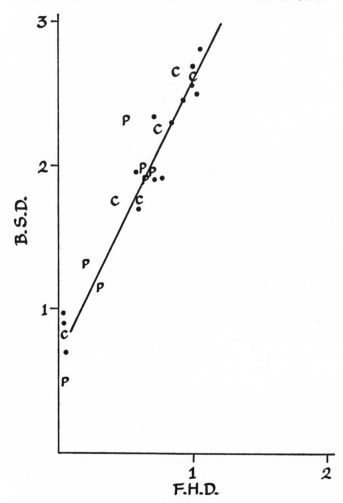

FIGURE 35. Same as Figure 34 except that now foliage height diversity for Puerto Rico is calculated from the *two layers* 0–2′, >2′ and that for the Canal Zone from the *four layers* 0–2′, 2′–10′, 10′–50′, and >50′. The correlation is much improved over that given in Figure 34 for three layers exclusively, indicating that diversity is based on a discrimination of fewer vegetation layers in Puerto Rico and more layers in the Canal Zone than is the case in the temperate United States. (After MacArthur, Recher, and Cody, 1966.)

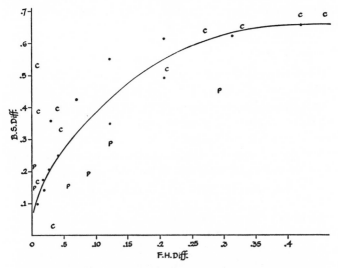

FIGURE 36. The index, BS diff., of bird species difference between habitats is plotted against the index, FH diff., between the same habitats. The conventions are the same as in Figures 34 and 35. The lower points for Puerto Rico (P) show that on this island the same degree of difference between habitats causes less difference between the bird faunas than is the case on the tropical and temperate mainlands. (After MacArthur, Recher, and Cody, 1966.)

horizontal line through either figure and see how many species it intersects. If food, as well as habitat, were restricted due to competition, then the number of species per patch of habitat should show a marked increase on the mainland. In addition, let us make the reasonable assumption that a layer in a homogeneous census area approximates a patch within which a species forages in a fine-grained fashion. Then the data in Figure 35, which show that a Puerto Rico layer has about as many species as one from Panama or temperate United States, also confirms that only habitat and not food within patches is compressed by competition. Since there are some unresolved difficulties in censusing small tropical areas, these data are suggestive but not conclusive. Further documentation would be highly

113

desirable, especially in species for which the separation of habitat and food is not so clear cut.

Ecological Shift

While expansion and contraction can be readily conceptualized and documented at this stage of the study, the other principal class of short-term change, ecological shift, is harder to prove in general terms. It is apparent on at least *a priori* grounds that in order to have the greatest chance for success the colonizing species should place itself so as to minimize competition or, actually, to maximize $K - \sum_i \alpha_i x_i$, where K is the carrying capacity of the environment in the absence of competitors, α_i is the competitive reduction in K by an individual of species i, and x_i is the number of individuals of species i. Both K and the combination of species i will differ between island and mainland and among islands. Those species will prove superior colonists that are flexible enough in behavior and growth habits to reduce $\Sigma \alpha_i x_i$ by insinuating themselves into portions of the environment unused by various, unpredictable sets of x_i. The relationship of the degree of ecological shift to plasticity and the competition coefficients α_i remain to be documented.

On the other hand, *evolutionary* shift in ecology is a very well-documented event. It will be treated in some detail in Chapter 7.

Habitat Islands on the Mainland

Many authors have pointed out that patches of habitat—a mountain top, a relict bog, or just a recently cut part of a forest—are islands. But these islands are different in that the space separating them is not barren of competitors. A true oceanic island clad in spruce forest would have few bird species present, but those would be typical of spruce forests; a small patch of spruce amid southern deciduous forests might be well populated mostly with deciduous forest species which had overflowed from adjacent habitats. On true islands, the immigration rate is so low that only species with positive r can be expected at any time, because, as we pointed out in Chapter 4, a species with birth rate, λ, slightly less than mortality, μ, may stay around for only about $1/\lambda \ln (2K - 1)$ years, where K is the carrying capacity of the habitat. Suppose our habitat has $K = 100$ so that $\ln (2K - 1)$ is about 5.3. Then a single pair of immigrants of a species with $\lambda = 3$ would normally last less than two years. Hence immigrants must come oftener than once every two years to maintain the population. On the mainland this is very easy, since many pairs of territorial species may have the alternatives of not breeding at all or of breeding in the adjacent, unsuitable habitat island. Such species will persistently overflow into the island and may even settle in such high density that they can out-compete the specialists in the island habitat. By the same token, because such habitat islands are likely to be partially occupied by the overflow from adjacent habitats, appropriate species find it harder to colonize than they would a true island.

Figure 37 shows another kind of biogeographic "island" on a continent. The numbers of breeding land bird species in the squares clearly increases where the physiography of the land is complex and increases toward the tropics. But the impoverishment of Florida, Baja California, and Yucatan can only be explained by recalling that the immigration rates onto these peninsulas is reduced by the absence of land around most of their boundaries that could serve as source regions. A similar peninsular effect has been demonstrated in mammals by Simpson (1964).

The Increase of Relative Diversity Through Filtering

Filtering is the reduction of numbers of species, genera, or higher categories during dispersal. It can be brought about by complete failure of certain taxa to disperse over the barrier; by the lack of suitable habitats for certain taxa on the recipient island; or by the impoverishment of the island biota through the distance effect. In the latter situation, the propagules of the missing taxa arrive on the island but at intervals longer than the average survival time of the populations they found. By definition, filtering reduces

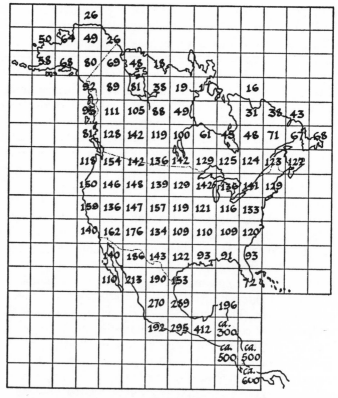

FIGURE 37. The numbers of breeding land bird species in squares roughly 300 miles across, taken from Robbins, Brunn, and Zim (1966) and, for Mexico, from Blake (1953).

the *absolute* diversity of colonizing taxa. We will now show how, on the other hand, filtering can increase the *relative* diversity. By this we mean that the number of genera and higher taxa per species is increased in the set of colonizing species.

Consider a group of genera consisting of species at least some of which occupy multiple habitats in source regions and stepping stones. If each habitat is equally suitable as a point of departure for overseas dispersal, and if the dispersal parameters do not vary as a function of the number

116

of species in the genus, then the percentage of expanding species per genus should not change with the size of the genus. In other words, genera containing many species should not differ in dispersal ability from genera containing few species. And the amount of relative diversity among expanding species should equal that among static species.

However, in the Indo-Australian ant fauna, this has proved not to be the case. As demonstrated in Figures 38 and 39, the smallest genera contain the highest percentage of expanding species. Furthermore, the number of expanding species per genus appears to have a firm upper limit. The following explanation has been offered by Wilson (1961). Although conclusive documentation is lacking, there appears to be no clear decline of average population density with increase in size of genus. This is not to say that such a relation cannot occur within the same genus in going from region to region, but only that it is not apparent in comparing different genera containing different numbers of species

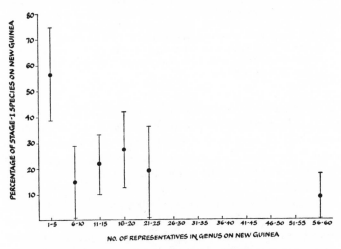

FIGURE 38. The relation of percentage of widespread, apparently expanding species ("Stage-1" species) to size of the genus on New Guinea, in the ant subfamilies Ponerinae, Cerapachyinae, and Myrmicinae. The frequency of all genera combined in each size class is given, along with the 95% confidence limits. (From Wilson, 1961.)

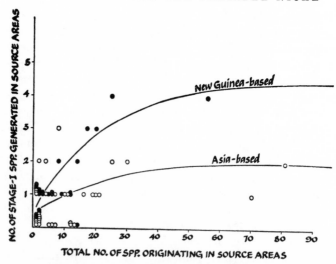

FIGURE 39. The relation of size of genus to number of wide-spread, apparently expanding species in the Ponerinae (including Cerapachyinae), and Myrmicinae in Asia (open circles) and in New Guinea (closed circles). Only those widespread species that occur on New Guinea were included in this tabulation, which accounts for the smaller number generated from Asia. Note that no genus, regardless of size, has been able to generate more than four expanding species from New Guinea or more than three from Asia that reach as far as New Guinea. (From Wilson, 1961.)

on New Guinea. Hence population size (K), the one parameter that might be expected to change with species packing, seems not to vary in this way from genus to genus on the same large island. But the habitats in which the ants occur do differ greatly in the percentages of expanding species they contain. Those habitats with the highest percentages are situated closer to the coast and rivers from which propagules are most likely to be launched overseas; they are almost certainly the principal "staging" areas for inter-island dispersal. The staging habitats contain relatively few species, and these represent the smallest genera to a disproportionate degree. A large amount of evidence has been accumulated that indicates the frequent occurrence of competitive exclusion among ant species on islands. The more similar the species, the more likely they are to exclude one another. Consequently there is a limit on the number of closely related (congeneric) species that can occur in the staging habitats at the same time. This in turn restricts the number of congeneric species that can expand simultaneously. Hence the staging habitats limit the transmission of within-genus diversity from island to island, while promoting diversity at generic and higher levels.

A parallel effect appears on small islands. Figure 40 shows

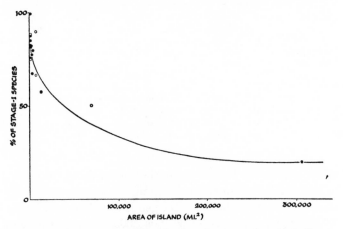

FIGURE 40. The percentage of expanding ponerine (including cerapachyine) ant species on islands of various areas in the Indo-Australian region. Open circles, Moluccas and Celebes; closed circles, Melanesia exclusive of Fiji. (From Wilson, 1961.)

how the percentage of expanding species increases with a decrease in island area. Stepping-stone islands are typically small. We will show in Chapter 6 that regardless of their smallness, such islands play a decisive role in dispersal under a wide range of conditions. It seems probable that they impose the same kind of filtering and enhance diversity in the same way as the staging habitats on larger, source regions.

What appears to be an extreme example of the same effect has been described by Gressitt (1964). On several subantarctic islands recently surveyed and on Antarctica itself, there is a total of less than 1,000 arthropod species. Yet these are relatively very diverse in a taxonomic sense—there is an average of less than two species per genus. The same high relative diversity occurs in the vascular plants, as seen in Table 11. The biota of New Zealand, which is the principal

TABLE 11. The average number of species per genus in the flora and insect fauna of New Zealand, three subantarctic islands, and Antarctica (based on data from Brown, 1893; Allan, 1961; and Gressitt, 1964).

	New Zealand	Auck- lands	Camp- bell	Mac- quarie	Ant- arctica
Arthropods	6.3[a]	1.3	1.4	1.6	1.5
Vascular plants	5.0	—	1.2	1.7	—

[a] Coleoptera only; complete data from other larger arthropod groups were not available. This figure (6.3) compares with the ratio 1.3 for the Coleoptera of Campbell Island and appears to be an indicator of the larger average size of arthropod genera in New Zealand.

source region for the subantarctic islands listed, has a distinctly higher average species number. The effect extends on to categories higher than the genus. On Campbell Island, for example, there are only about 300 arthropod species, but these represent no less than 105 families.

Finally, the same effect has recently been thoroughly documented and analyzed in the bird faunas of small islands by Grant (1966), who presents still other data to relate the phenomenon to interspecific competition.

The principle of increase of relative diversity by filtering, now that it has been shown to be of general occurrence, has several important consequences for biogeographic theory. For example, it means that newly colonized islands start with a greater number of stocks than would be anticipated from an inspection of a random sample of species in the source region. We can also conclude that those species that are ecologically the most distinct (as a concomitant of their membership in distinct, smaller genera) have an advantage in dispersal. Another implication is that species turnover is apt to be greatest in the staging habitats and stepping stones, due to a higher competitive pressure and replacement rate by newly expanding species. Also, the role of islands as stepping stones and their filtering effect increase as the percentage of the island area covered by the staging habitats. These aspects of biogeographic theory have not been quantified and are so far supported only by limited documentation. Among the unanswered questions is the following: Is the high percentage of expanding species on small islands, and conversely the low percentage of endemics, due to a higher preponderance of staging habitats on these islands; or is it due simply to a higher turnover rate caused by smaller population size? Perhaps both factors contribute, to a relative degree yet to be determined.

SUMMARY

There is a limit to the number of species persisting on a given island. An island is closed to a particular species either when the species is excluded by competitors already in residence or else when its population size is held so low that extinction occurs much more frequently than immigration. In discussing both the theory and exemplification of competitive replacement, we showed that species can be excluded from an island in many and unexpected ways. This helps explain in a very general way why so many species are absent from places where the environment seems superficially well suited for them. A closer examination of the composition and behavior of resident species should often reveal the causes of exclusion, so that random processes in colonization need not be invoked.

When a species invades a new island, it encounters in almost every case an environment that is different to some degree. Most frequently the change is biotic: the island contains new combinations if not new kinds of predators, prey, and competitors. There is a tendency, by no means universal, for a colonizing species to respond by either contracting or expanding its niche. As a rule, it contracts on meeting more competitors and expands on meeting fewer of them. A priori considerations (supported by limited evidence in birds) suggest that the compression occurs much more readily in habitat than in diet.

"Habitat islands," i.e., patches of habitat surrounded by other habitats that are distinct but not radically different, differ from complete islands, such as land surrounded by water or water surrounded by land, in at least one important respect affecting species composition. Species already present on the habitat islands are faced with constant pressure of high immigration of less-well-adapted species drawn from the surrounding habitats. Simultaneously, species attempting to colonize a habitat island should find it harder to do so because of the greater diversity of competitors opposing them at any given moment.

During dispersal along island chains, there is a loss in the absolute numbers of both species and higher taxa. But as this filtering proceeds in an absolute sense, the genus/ species ratio can increase due to competitive replacement in the more restrictive island habitats. The result, which has been documented in both plants and animals, is an increase in relative diversity in the species that succeed in dispersal.

CHAPTER SIX

Stepping Stones and Biotic Exchange

Stepping-stone Dispersal

A central but hitherto intractable problem of biogeography is the estimation of the rate of exchange of species among source areas. This can be alternatively described as the estimation of the relative contributions of species of differing source areas to a recipient area. Although complex, the subject can be treated with enough formal analysis at this time to yield some qualitative generalizations.

Let us begin with the simple and special case of competition between a "major" source island and a smaller, intermediately positioned "stepping-stone" island. The following question often arises in practice: By how much does a stepping stone that has been colonized add to the overall dissemination rate to a third, farther island? The problem can be restated in a more manageable form as follows: If a propagule of a given species arrived on the third (recipient) island, what is the probability that it came from the stepping stone rather than the more distant source island? No direct measurements have been made of such competition between islands. But theoretical solutions of certain simple cases are possible. Consider, for example, the case of three islands arranged in a straight line, in a region of uniform wind and ocean current patterns (Figure 41).

The following measurements are required for the construction of the simplest conceivable model:

w_1, width of the "stepping-stone" island.

w_2, width of the more distant "source" island.

w_r, width of the recipient island measured at a right angle to the line joining the three islands.

d_1, distance from the stepping stone to the recipient island.

d_2, distance from the source island to the recipient island.

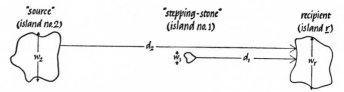

FIGURE 41. Diagram defining the measurements used in the stepping-stone model.

We must first consider the relation between the width and area of an island and the number of propagules leaving its shores. If the average dispersal distance of the propagules is a large fraction of the width of the island, then the number of propagules leaving the island's shores in a given period of time can be approximated as a simple linear function of the island's area or as a square function of the island's diameter. In the model of Figure 41, where the islands are roughly circular in outline, it is αw_i^2, where α is a constant. If on the other hand the average dispersal distance of the propagules is very small compared to the width of the island, the number of propagules leaving the island's shores in a given period will be a linear function of the circumference of the island, that is, βw_i, where β is a constant. In cases where the average dispersal distance is intermediate relative to the island width, the power of the function will have an intermediate value between unity and two.

For convenience we will take the first case, in which the numbers of propagules leaving the source and stepping stone are αw_2^2 and αw_1^2 respectively. Other area-propagule functions would give the same qualitative results.

Of course most of the αw_i^2 propagules departing overseas will perish at sea (or in whatever inhospitable environment separates the ecological "islands" to which the theory applies). The number reaching a given distance d is determined by a probability distribution which, in a given set of environmental conditions, is species-specific. There is reason to believe, on the basis of existing evidence, that the *form* of this survivorship probability distribution, quite in addition to the mean dispersal distance, varies greatly among larger taxa of plants and animals. Such variation can have a marked effect on the patterns of dispersal. For example, consider differences that would emerge among the simplest classes of distributions: the uniform, the exponential, and the normal.

A *uniform distribution* will be approached, although never wholly attained, if the mean dispersal distance is relatively very great, at least very much greater than d_2 in Figure 41. In this case the probability that a given propagule arriving on the recipient island came from the stepping stone rather than the source island would be w_1^2/w_2^2, the ratio of the areas of the stepping stone and source island, or, in fact, the ratio of the numbers *leaving* the stepping stone and source. A vertebrate which only departs when it can see an island may be a case. As a general rule, an organism x miles above the ocean can see an island y miles in elevation from a distance of $89(\sqrt{x} + \sqrt{y})$ miles. Thus a bird one mile high can see an island two miles high from about $89(1 + 1.4) = 214$ miles. The species might have uniform dispersal over this distance.

An *exponential distribution* will result if the propagule moves in a constant direction and if the probability that in a fixed interval of time it will cease moving, e.g., that it will fall from the air to the ground or sea, remains constant. Such would likely be the case for passive terrestrial propagules carried over the sea in a steady wind or in a steadily moving cyclone. The number of propagules surviving after travelling the distance from one of the source islands to the recipient island will be very roughly

$$\alpha w_i^2 e^{-d_i/\Lambda},$$

where Λ is the mean dispersal distance. Exponential dispersal might prove to be common if not universal in plants and insects that disseminate propagules passively through the air. A probable example is given in Figure 42.

To complete the model for intersource competition in a taxon with exponential dispersal, we should further take into account the relative probabilities of propagules taking the right direction to the recipient island. From Figure 43,

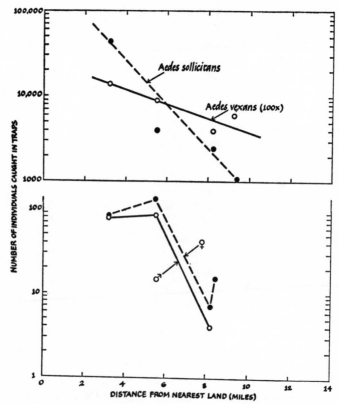

FIGURE 42. Over-water dispersal of two groups of Diptera known to be relatively vagile, based on collections made at buoys and lighthouses in Delaware Bay. The sample points were surrounded by land on all sides, and hence the curves are subject to distortion due to convergence of individuals from various directions. *Above.* Possibly exponentially distributed dispersal of two species of *Aedes* mosquitoes. *A. vexans* has fewer individuals but a greater mean dispersal distance. This form of dispersal curve is consistant with the known tendency of mosquitoes to be carried long distances by wind. *Below.* A possible case of normally distributed dispersal in horseflies (Tabanidae). A tendency toward this form of dispersal would be consistent with the fact that tabanids are exceptionally strong flyers. Not corrected for the inverse-square decrease expected from the increase of sample area with distance. Data of D. MacCreary and L. A. Stearns, in Wolfenbarger (1946).

126

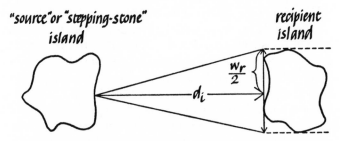

FIGURE 43. Diagram illustrating the angle embracing correct directions that can be taken by propagules.

it can be seen that the probability that the right direction was taken by a propagule travelling as far as the recipient island is the ratio of the angle subtended by the opposing diameter of the recipient to 360°. Thus in the simplest possible case, the total number of propagules n_i reaching the recipient island can be approximated as

$$\frac{2 \tan^{-1} (w_r/2d_i)}{360°} \; \alpha w_i^2 e^{-d_i/\Lambda}$$

In actual estimates, it would be better to substitute the true area of the island for w_i^2, since islands are seldom if ever neatly circular. It should be clear by inspection that the most important term by far is d_i/Λ; in other words, the degree of isolation relative to the mean dispersal distance will usually be decisive. In fact, the directional component is so minor and also so subject to variation due to unknown factors (such as orienting ability of the propagules) that it is perhaps misleading to guess at its explicit form. We have used the simplest possible term here, $\tan^{-1} (w_r/2d_i)/360°$, only as a device to permit us to make explicit, experimental computations.

Returning to the original problem, we are interested in the probability that a given propagule landing on the recipient island comes from the stepping-stone island (no. 1) rather than the source island (no. 2). This is simply the fraction of all the propagules reaching the recipient island that came from the stepping-stone, or $n_1/(n_1 + n_2)$.

A third kind of distribution that might be approached

127

under certain simple but different conditions is the *normal distribution*. It can be expected where propagules fly actively through the air on a randomly changing course, thus acting like particles in a diffusing gas; where they fly on a set course for a period of time that is normally distributed; or where they are borne on a sea-going "raft," such as a floating log, which has a normally distributed persistence time. The fraction of individuals continuing with increase in distance falls off at the rate of e^{-x^2} rather than e^{-x}, the term in exponential dispersal. An apparent example of a dispersal curve approaching a normal distribution was given in Figure 42. In the normally distributed case we can estimate the number of propagules passing from one island to another over a distance d_i as

$$n_i = \tan^{-1} (w_r/2d_i)\alpha w_i{}^2 \left[1 - \frac{2}{\sqrt{2\pi}} \int_0^{d_i} e^{-x^2/2}\, \mathrm{d}x \right]$$

where d is measured in units of the standard deviation σ. The ratio $n_1/(n_1 + n_2)$ can be calculated directly from this expression. For simplicity the bracketed term in the formula can be approximated by the expression

$$\sqrt{\frac{2}{\pi}}\, \frac{e^{-d_i{}^2/2}}{d_i},$$

since the term is simply the number of individuals that reach the distance d_i or beyond, where d (the distance travelled) is a normally distributed random variable. As in the exponential model, the directional component is apt to be minor; it has been included in explicit form only to permit computation.

Notice that since we are dealing with ratios, we do not need to know the absolute rate of propagation of propagules from the source and stepping-stone islands; in other words α cancels out. Similarly, so long as the three islands are aligned, the effects of variation in wind and current direction and in velocity are minimized. Such variation can be entered into the equations if known, but for our present purpose they can be disregarded.

Elementary models constructed from the foregoing relationships can be employed to make certain new qualitative inferences that will hold over a wide range of conditions. To this end isoclines of $n_1/(n_1 + n_2)$, the relative contribution of the stepping stone, for the exponentially distributed case, with $w_r = w_1 = d_1 = \Lambda$, are given in Figure 44. The same isoclines for the normally distributed case, with

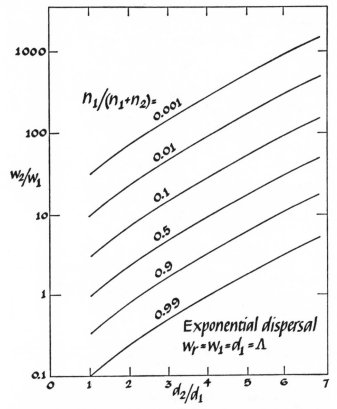

FIGURE 44. Purely exponential dispersal involving the simplest and most regular conceivable participation by a stepping stone. Isoclines are given for the probabilities that a given propagule arriving on the recipient island came from the stepping stone (island 1) rather than the source island (island 2), in other words $n_1/(n_1 + n_2)$ in the model. They show the rapidly increasing importance of stepping stones of almost all sizes as their relative distance to the recipient island is decreased.

$w_r = w_1 = d_1 = \sigma$, are given in Figure 45. Since σ of the normal distribution is equal to 1.253 times the mean deviation from the mean (MD), the equivalent of Λ in the expo-

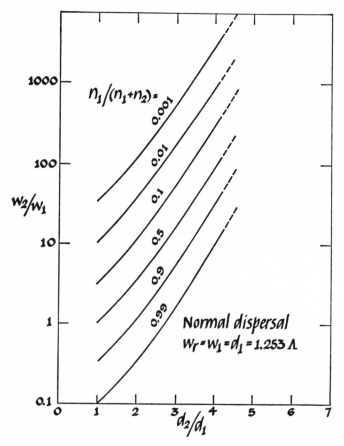

FIGURE 45. Purely normal dispersal. Conventions as in Figure 44. The curves show the even greater importance of stepping stones in this second class of dispersal distributions.

nential case, the conditions for Figures 44 and 45 are closely comparable. They apply to two sets of propagules with approximately the same average dispersal distance in overseas dispersal. Figure 46 shows the result of decreasing the

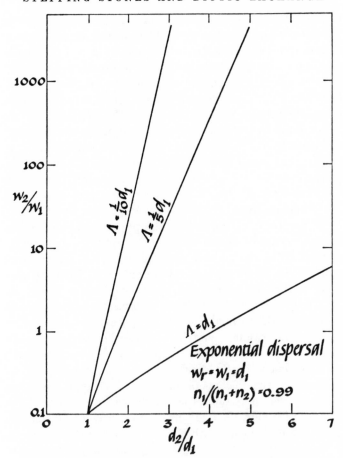

Figure 46. Purely exponential dispersal, conventions as in Figure 44. The 0.99 isocline is shown as it shifts rapidly upward with a decrease in the mean dispersal distance (Λ) of propagules. If Λ is a small fraction of the width of the gaps separating the islands, the relative contribution of stepping stones of almost all sizes is overwhelming.

mean dispersal distance in the exponential case. Similar results are obtained for the normally distributed case.

We do not believe that many real cases will prove to be truly exponential or truly normal. Perhaps none are. It is more likely that real dispersal distributions are com-

131

pounded from simpler distributions, or are complex due to the presence of other, as yet unidentified features. Given a certain mean dispersal distance, densities of real propagule populations probably decline at least as rapidly as predicted by the purely exponential distribution. In other words, estimates from the exponential distribution, which assumes a strictly passive role of the propagule, probably set an upper limit for propagule densities at moderate to great distances from the source regions. At the same time, it is known that many animal species do disperse outward by active movements, continuing in motion until they reach a resting place or exhaust their energy reserves (Cockbain, 1961; Sheppe, 1965). It is also generally true that in most or all species of flying insects the pre-reproductive adults engage in special dispersal flights, during which they fly upward from the ground and are carried along in large numbers by wind currents (Johnson, 1957, 1963). Under such conditions, propagules flying over water would be expected to disperse until their reserves are exhausted, at which time they would sink rather quickly to their deaths. The general result would be a dispersal distribution approaching the normal. If an additional factor, such as death or a tendency to fly downward, began to act prior to exhaustion, the decline of propagules at great distances could be even more rapid than predicted from a normal distribution (Taylor, 1965). There is apparently no information available at this time with which to evaluate the matter in a quantitative fashion. By taking two simple but quite different dispersal curves, we will at least be able to show that variation in the form of dispersal curves is capable of creating important variation in the distribution patterns.

Examination of the curves in Figures 44 and 45 shows, for example, that where the mean dispersal distance is the same, stepping stones are far more important for species with normally distributed dispersal than for those with exponentially distributed dispersal. Translated into biological terms, it can be expected that stepping stones are more important to species whose propagules tend to disperse actively or on floating "rafts," such as birds, mammals, and some plants and arthropods. They are relatively less important to species whose propagules tend to be dispersed passively in the wind, such as most microorganisms and many higher plants and arthropods.

The potential significance of the distinction can further be illustrated by imaginary examples in which the parameters are varied widely. One that involves competition between the biotas of two large islands on continents is given in Figure 47. Data are as yet lacking to test the validity of this predicted effect, but it seems intuitively likely that such a phenomenon exists and has weight in real biotic exchanges. In short, the *form* of the dispersal curve, as well as the dispersal power of the species, probably influences biogeographic patterns.

The curves of Figures 44–46 illustrate the principle that the role of stepping stones in colonization increases very rapidly as the dispersal power decreases. The effect can easily become total for even the largest source biotas. For example, if the mean dispersal distance of propagules is as little as one-tenth the distance to the stepping stone, and the stepping stone is located as little as two-thirds the distance to the recipient island, almost all of the species capable of prolonged existence on the stepping stone will reach the recipient island from it rather than from the ultimate source island or continent. When we insert the known mean overland dispersal distances of various species of plants and animals, we seem to be led to the surprising conclusion that dispersal across gaps of more than a few kilometers is by stepping stones wherever habitable stepping stones of even the smallest size exist.

A case in point is Futuna and the Wallis Islands, which lie midway between Fiji and Samoa and therefore serve as potential stepping stones for species pressing outward from Melanesia to Polynesia (Figure 48). The ant fauna of Futuna and the Wallis Islands is merely an extension of the fauna of eastern Melanesia, particularly Fiji; there are no known endemic species (Wilson and Hunt, 1967). Samoa and the remainder of western Polynesia are also populated by native species which are drawn from eastern Melanesia.

133

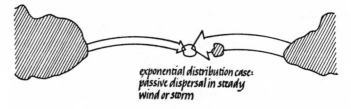

exponential distribution case:
passive dispersal in steady
wind or storm

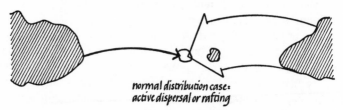

normal distribution case:
active dispersal or rafting

FIGURE 47. An imaginary case involving the biotas of two source islands competing to colonize a recipient island. The recipient island is set midway between the source islands, and the stepping stone set one-fourth the distance from the recipient island to the source island on the right. The width of the source island is made equal to Λ and the width of the recipient island and one-tenth the width of either source island. The widths of the source islands are set equal, and neither source island is given an advantage in the departing wind or water currents. The stepping stone is now colonized by propagules from the source island on the right. From that time, the biota from the right source island contributes twice as many propagules to the recipient island as the left source island in the case of species that disperse exponentially, but one hundred times as many in the case of species that disperse normally.

Discounting the endemics, which are ultimately Indo-Australian (and probably Melanesian) in origin, no less than 15 of the 26 native Samoan species are also found on Futuna or the Wallis Islands or both. Thus Futuna-Wallis has the essential species pool. The probabilities that propagules arriving on Samoa come from Futuna-Wallis rather than Fiji—in cases where the species occur on both Futuna-

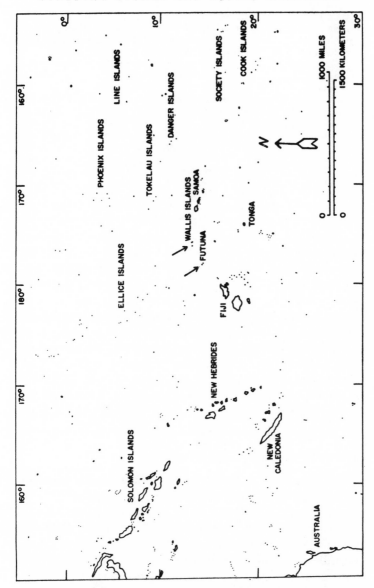

FIGURE 48. Map showing the location of Futuna and the Wallis Islands, potential stepping stones between Melanesia (including Fiji) and Polynesia (including Samoa).

TABLE 12. The probability $n_1/(n_1 + n_2)$ that a propagule arriving on the Samoan island of Savai'i came from Futuna (or Wallis) rather than from Fiji, assuming one or the other of two dispersal curves and a wide range of mean dispersal distances (Λ) for various species.

Mean dispersal distance in miles (Λ)	EXPONENTIAL DISPERSAL		NORMAL DISPERSAL	
	Futuna	Wallis	Futuna	Wallis
0.01	>0.999999	>0.999999	>0.999999	>0.999999
0.1	>0.999999	>0.999999	>0.999999	>0.999999
1	>0.999999	>0.999999	>0.999999	>0.999999
10	0.999993	>0.999999	>0.999999	>0.999999
100	0.075971	0.173986	0.455888	0.968234
1000	0.019107	0.016662	0.009369	0.030721
10000	0.016587	0.012998	0.008173	0.025221

Wallis and Fiji—have been calculated from the stepping-stone models and are given in Table 12. They suggest that for most species, Futuna and the Wallis Islands, together with a few other small intermediate islands such as Niuafo'ou, Tafahi, and Niuatoputapu, serve as the source regions for Samoa and other parts of Polynesia. Note that essentially the same result is obtained from both the exponential and normal distributions. The important point is that almost any conceivable dispersal distribution, in which the mean dispersal distance of propagules is much less than the width of the water gaps separating the islands, would have the same qualitative outcome.

The Fringing Archipelago Effect

An extension of the same elementary models leads to other, even more unexpected results. One such result concerning biotic exchange can be stated simply as follows. If a single stepping stone appears, it should not alter the proportion of species exchanged between the source areas, regardless of its location. But if two or more stepping stones appear in succession and are colonized one after the other (the much more typical case in nature), together they will increase the relative flow out from the source area closest to them.

The argument leading to this inference can be illustrated with a simple imaginary numerical example. Suppose, as in Figure 49 (upper), the two source areas (A and B) are comparable in area and hold different biotas that exchange species at the same low, fixed rate. For convenience in computation this rate is taken as unity. If a new island (R_1) of equal size were to emerge between them, it would be populated by species from A and B roughly in proportion to the relative flow of propagules from A and B. Suppose R_1

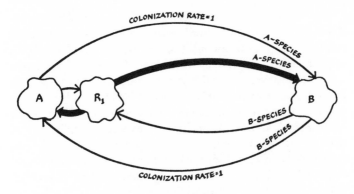

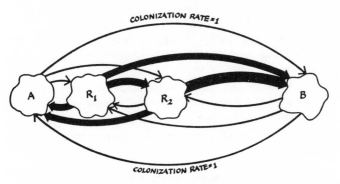

FIGURE 49. *Above.* Where only one stepping stone (R_1) exists, it contributes equally to the exchange of species between A and B, whatever its position, and neither A nor B gains an advantage. *Below.* When a second stepping stone (R_2) arises and is nearer A, i.e., A acquires a fringing archipelago, A gains the advantage.

137

to be 1 unit distance from A and 3 units from B and the results from the exponential model given in Figure 44 applied. We would expect a flow of 49 from A and 1.1 from B, resulting in 0.978 of the species on R_1 originating from A and 0.022 originating from B. At equilibrium the total flow from R_1 to A would be 49 and from R_1 to B would be 1.1. However, the *flow of A species* from R_1 to B would be 0.978 × 1.1 = 1.08, and similarly the *flow of B species* from R_1 to A would be 0.22 × 49 = 1.08. Thus the exchange would be equal.

Now consider the case given in the lower half of Figure 49. After R_1 has reached equilibrium (and it is an essential condition of the argument that R_1 be colonized first), R_2 appears midway between A and B. In spite of its intermediate position, R_2 will not acquire equal sets of species from A and B respectively. Many of its propagules (in this particular case most of them) will come from R_1, with a few each from A and B, resulting in an eventual equilibrium of 0.93 species of A origin and 0.07 species of B origin. The feedback of species from both stepping stones to the source areas will thereafter consist predominantly of A species. In our special case, the total flow of A species to B will have increased from 1, when no stepping stones existed, to 4.7, when two existed; while the total flow of B species to A will have correspondingly increased from 1 to only 2.2.

The same qualitative result—a favoring of continents with fringing archipelagos in species exchanges—seems intuitively to hold for both the exponential and normal dispersal models and for all shapes and sizes of archipelagos. If, as we have reasoned, stepping stones are of major importance in dispersal in most imaginable cases, this effect should be of some general importance in biotic exchange.

The Problem of Radically Mixed Biotas

Of the many unsolved problems of comparative biogeography, some of the most intriguing concern the biotas of radically mixed origin. The Florida Keys, for example, are occupied by higher plants and insects of mostly Antillean origin and a vertebrate fauna of almost wholly North American origin (Small, 1933; Duellman and Schwartz, 1958). Hawaii has a predominantly North American bird fauna and a Polynesian flora and invertebrate fauna (Mayr, 1943; Zimmerman, 1948; Cooke and Kondo, 1960). Even more striking is New Guinea, with primarily Oriental insects and higher plants and Australian vertebrates (Mayr, 1951; Gressitt, 1961). How can these differences in direction of flow among the major taxa be explained? By using New Guinea as an example, it should now be possible, if not to provide a final answer, at least to define the problem more sharply, to examine the potential usefulness of the dispersal theory just presented, and to identify the important weaknesses in our empirical information.

It might at first thought seem easy to account for New Guinea's biotic chimaera as the outcome of differences in the number of propagules disseminated by the various taxa. Perhaps plant and insect propagules are so numerous that Asiatic elements are able to cancel out the advantage of greater proximity enjoyed by the Australian elements. But this logic would be flawed, for even if insects and plants produced vastly greater numbers of propagules than vertebrates, the ratios of propagules of Asian versus Australian origin could not change in any simple way that would tip the balance for one group but not the other.

The differences could be explained if vertebrate propagules moved predominantly northward and westward, while insect and plant propagules moved predominantly southward and eastward. Such an appealing dichotomy seems very unlikely to exist in fact. Water currents, which are apt to favor vertebrates, move in the wrong direction—southward and eastward. Prevailing winds, which on the other hand are more likely to carry insects, also move in the wrong direction—northward and westward. Plants are carried by wind, water currents, and birds and should therefore be brought in from various directions. Cyclones are rare along the equator area but generally move southward from the vicinity of New Guinea (Wiens, 1962). Thus only a weak correlation exists, and it is the reverse of the one expected.

The differences could be explained if the water barriers separating New Guinea (together with Australia) from the Sunda Shelf allowed no vertebrates to cross. This appears to have been the case only in the fresh-water fishes of Cenozoic origin (Darlington, 1957).

The differences could be explained if insects and plants invaded New Guinea primarily at a time when the island was open to Asia, and the vertebrates invaded at a time when the island was more accessible to Australia, in particular during the life of the Sahul land bridge in the Pleistocene. But insects, plants, and vertebrates as a whole have evidently been dispersing and evolving in the Papuan region throughout at least the Cenozoic down to the present time.

The differences could be explained if the groups differed in their ability to penetrate new climates. Mayr (1951) expressed this hypothesis as follows: "The primary factor is that plants have very high dispersal facilities but are ecologically very demanding. They spread rather easily through the ecologically fairly uniform tropical belt from Malaya to New Guinea and the Solomon Islands but have great difficulty in becoming established in the rather arid sub-tropical zone of Australia. Animals, on the other hand, and warm-blooded vertebrates, in particular, have much greater dispersal difficulties, but once they have jumped an oceanic barrier they have apparently less difficulty in adjusting themselves to a new environment . . . . The sharp division between the Indian and the Australian fauna, as shown by birds, mammals, and other groups of animals, is an indication of the long-standing separation of the Asiatic Sunda Shelf and the Australian Sahul Shelf. The similarity of the floras in the tropical belt is an indication of the similarity of climatic condition." This explanation does not fit the insects, which are now known to consist of both eurytopic and stenotopic elements of mostly Oriental origin, nor does it apply to the cold-blooded vertebrates, which tend to be stenotopic yet of Australian origin.

One or more of these four injured hypotheses may yet be revived with new information. But for the moment, it is necessary to continue the search for explanation. We can conceive of three additional hypotheses that are not directly opposed by the evidence. All involve the dispersal or demographic properties described in our earlier models. They are not mutually exclusive. The first hypothesis, illustrated in Figure 50A, states that the different invasion patterns are an outcome of a difference in the *mean dispersal distance* (Λ) of propagules. (Note that differences in absolute numbers of propagules, or αw_i^2 in the models, has already been ruled out as a unique explanation.) In order for this explanation to work, it is necessary that the mean dispersal distance be greater in insects and higher plants than it is in vertebrates. At first glance the evidence, such as it is, seems to indicate just the opposite. From published data of overland dispersal of plants and insects (e.g., Wolfenbarger, 1946), Λ is usually to be calculated in the tens or hundreds of meters or

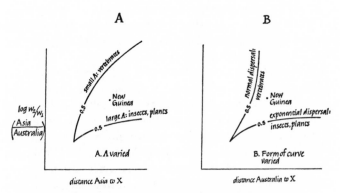

FIGURE 50A & B. Two models that explain the mixed origin of the New Guinea biota as an outcome of variation in dispersal parameters. Log w_2/w_1 measures the ratio of propagules disseminated from Asia and Australia respectively. The 0.5 isocline is the line below which propagules arriving in New Guinea are predominantly Australian; above the line they are predominantly Asian. In A, vertebrates are assigned a smaller mean dispersal distance (Λ) than insects and plants, and New Guinea is placed between the two 0.5 isoclines. The form of the dispersal curve is not varied decisively. In B, the two groups are given a different form of dispersal curve, while Λ is not varied decisively. The same result can be obtained by combining decisive changes in both parameters.

141

less; those of vertebrates are almost certainly greater. But it would be quite wrong to exclude the hypothesis on the basis of this evidence. The reason is that we have almost no quantitative information on the dispersal of organisms over water or inhospitable environments generally. Recent studies on insects (e.g., Johnson, 1957; French and White, 1960; Johnson et al., 1962; Hurst, 1964; French, 1965) have revealed a surprising ability of the adults of many flying species to climb high into the air and be carried tens or hundreds of miles during the "migratory" period of their life. These windborne propagules survive well even at 5,000 feet altitude or higher (Taylor, 1960). We also lack measurements of dispersal during storms, which cause Λ values of most species to increase. On such occasions the Λ values of plant and insect species are likely to be increased more than those of vertebrate species, first because they are more passive and second because they are more likely to survive rough passage by storm winds or heavy seas. Yet while we can guess that Λ values increase differentially between the two groups, there is no way of knowing whether they increase enough to tip the balance.

The second hypothesis, illustrated in Figures 50B and 51, explains the difference in invasion patterns as an outcome of difference in dispersal probability distributions. Here the results seem a priori fully consistent. Even if vertebrate curves tended slightly more toward normal distributions than plant and insect curves, as we have predicted they do, different invasion patterns could result. There are no opposing data—there are as yet almost no data of any kind. The shapes of dispersal curves will be harder to estimate than Λ values, especially if they prove to be complex in form, but they may be just as illuminating.

The third hypothesis is based on the observation, made in Chapter 4, that the average survival time of founding populations increases exponentially with r, the intrinsic rate of population increase. If species have a high r, a small number of propagules have a high probability of "taking" on the invaded island; and an increase in the number of propagules will not augment this probability by very much. Plants and

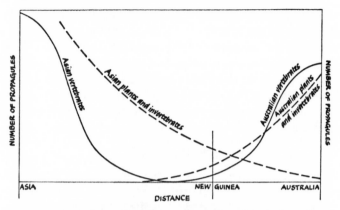

FIGURE 51. An extension of figure 50B, illustrating the hypothesis that differences in affinities can be explained by variation in the form of the dispersal curves. In this example New Guinea, which is closer to Australia than to Asia, nevertheless receives more plant and invertebrate propagules of Asian origin because the exponential form of their dispersal permits the greater absolute number of Asian propagules to outweigh the disadvantage in distance. On the other hand, fewer vertebrate propagules of Asian origin reach New Guinea because the normal form of their dispersal does not permit the disadvantage in distance to be overcome.

insects characteristically have high r values, perhaps enough so that the greater distance from Asia will not overcome the effects of Asia's greater species diversity and higher production of propagules. More Asian species will land propagules on New Guinea, because there are more in the Asian species pool than in the Australian species pool. There may be fewer propagules per species, due to the distance effect, but the survival rate per species may be nearly as high due to the r effect. The result could be a predominance of Asian species on New Guinea. Vertebrates on the other hand have a low r. The fewer propagules (per species) arriving from Asia will therefore give rise to populations with significantly shorter average life spans than the Australian immigrants, and the initial advantage conveyed by a larger species pool will be lost. The result could be a predominance of Australian species on New Guinea.

There are thus at least seven possible explanations of

radically mixed biotas, no two of which are mutually exclusive. In the case of New Guinea, the evidence seems to exclude four of them. The surviving three hypotheses involve variation in dispersal and demographic parameters, the consequences of which have been delineated by the use of deductive models. They can be tested only by the determination of dispersal curves over water under various weather conditions.

SUMMARY

Since measurements of overseas dispersal are at present almost non-existent and will be difficult to obtain in the future, we have attempted to explore the subject of biotic exchange by means of dispersal models.

The potential role of stepping-stone islands in increasing dispersal has been emphasized by the results of this study. It appears that even minute islands can significantly enhance biotic exchange providing they are able to support populations of the species in the first place. If they are relatively large and close to the recipient island, they can increase the flow of propagules by many orders of magnitude.

The form of the dispersal curve, as well as the average distance travelled by propagules, can affect dispersal patterns. Slight differences among taxa of plants and animals in these two properties can result in the formation of biotas of radically mixed origin on recipient islands. The case of the mixed biota of New Guinea, for example, might be explained in this fashion, although other explanations have not been discounted.

When a stepping stone arises between two source regions, it will increase the biotic exchange between them. But no matter how large it is or where it is located, it will not alter the original ratio of exchange. However, if a second stepping stone arises some time after the first one has been colonized, the source region closest to the two stepping stones will be favored in the exchange thereafter. If still more islands appear near the source region, the balance will be shifted further. This surprising theoretical effect has not yet been tested empirically.

CHAPTER SEVEN

Evolutionary Changes Following Colonization

Modes of Natural Selection on Islands

Since we believe that evolution through natural selection has produced the biotic differences which characterize islands, it is appropriate for us to study how natural selection acts on islands and, in particular, how it acts *differently* on islands as opposed to mainlands.

A slight digression on the history of natural selection theory will help put the discussion in perspective. As students are quick to point out to their teachers, the argument of natural selection is very nearly circular. In its circular form it says: (1) The more fit genotypes leave more descendants which, because of heredity, resemble their ancestors. (2) "Fit" genotypes are those which leave more descendants. Clearly part (2), which defines fitness, must be replaced by some other definition, or else the argument will be circular. Actually, many naturalists prefer to keep the argument circular and hence obviously true. They would say that the important point is that, empirically, some genotypes are more fit (i.e., leave more descendants) than others. Viewed this way, natural selection is not a mechanism but rather a restatement of a history. The restatement has no predictive power; we must wait and see which genotype leaves most descendants and then we will know that it was favored by natural selection.

In the 1920's and 1930's Fisher, Haldane, and Wright sought to remove the circularity and define fitness in terms of measurements which could be made on a present-day population. Lotka had already pointed out that a population whose age-specific birth and mortality rates are constant will come to expand exponentially with a rate r which then represents the difference between mean birth and death rates. For populations expanding with constant birth and death rates, r, or some equivalent measure (Fisher used r; Haldane and Wright used e^r which Wright called W) is then an appropriate definition of fitness. Clearly genes with greater r grow in population faster than those with lesser r, and they eventually come to predominate. The circularity is avoided because present-day instantaneous rates are used to predict future numbers of descendants. This is no longer circular because, in fact, it can even be wrong! For if the environment changes in ten years, present-day r values are not reliable predictors of numbers of descendants twenty years from now. However, for unchanging environments natural selection is made predictive.

Clearly we can never actually predict what genotypes will prove successful unless we can also predict what environmental changes will take place. One kind of environmental change is quite predictable, however (and others are at least statistically describable in terms of means, variances, autocorrelation, etc.). The predictable change is that which is caused by the expansion of the population, i.e., as a consequence of crowding. Crowding can affect populations in many ways, and each way affects which genotypes will persist into the future. To the ecologist, the most natural way to define fitness in a crowded population is by the carrying capacity of the environment, K, as follows (Mac-Arthur, 1962). Let the actual populations, n_1 and n_2, of two alleles be governed by the equations

$$\frac{dn_1}{dt} = f(n_1, n_2)$$
$$\frac{dn_2}{dt} = g(n_1, n_2)$$

(Notice that time does not enter explicitly into the right-hand sides, so we are assuming that the only environmental changes are due to the crowding itself.) Since crowding is assumed to be detrimental, there is some set of values of n_1 and n_2 so large that $dn_1/dt < 0$ and there will be a solution $dn_1/dt = f(n_1, n_2) = 0$ marking the inner boundary of this region. Within $f(n_1,n_2) = 0$, $dn_1/dt > 0$ and allele n_1 can increase. Similarly, within another curve $g(n_1,n_2) = 0$, $dn_2/dt > 0$ and allele n_2 increases. Figure 52 shows the four different ways in which the curves $f(n_1,n_2) = 0$ and $g(n_1,n_2) = 0$ can be related. The arrows on the curves show how the numbers of the two genotypes must change. For instance, on the line $f(n_1,n_2) = 0$ we know that $dn_1/dt = 0$ so that allele n_1 is neither increasing nor decreasing. Portions of this curve which lie inside the curve $g(n_1,n_2) = 0$ will then have n_2 increasing and n_1 staying constant (a vertical arrow directed upward) and portions outside of $g(n_1,n_2) = 0$ will have n_2 decreasing and n_1 constant. From the direction of the arrows it is clear that in part A of Figure 52 allele 1 will out-compete allele 2; in part B allele 2 will out-compete

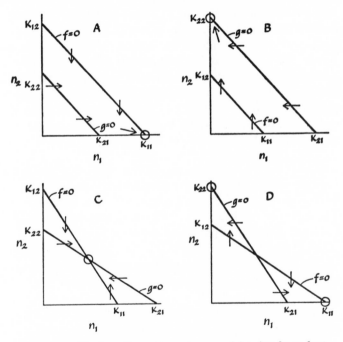

FIGURE 52. The four alternative outcomes of density-dependent selection of two alleles. The equilibrium points are encircled. Although the curves f and g are pictured as straight, they will probably often bend somewhat. See text for meaning of K's.

allele 1; in part C both will coexist stably; and in part D whichever has the initial advantage will win. We will now relate these graphs to ecology and genetics.

Where $f(n_1,n_2) = 0$ intersects the n_1 coordinate we have the value, K_{11}, such that $f(K_{11},0) = 0$. At that point the population consists of allele 1 only (hence all are homozygous) and K_{11} represents the number of these homozygous alleles which can just maintain themselves in this environment. K_{11} is the "carrying capacity" of the environment for these homozygous alleles. Where $f(n_1,n_2) = 0$ intersects the n_2 coordinate we have the number K_{12} such that $f(0,K_{12}) = 0$. It is the number of allele 2 which is just sufficient to keep allele 1 from increasing. Right along the n_2 coordinate there are no 1 alleles, of course, but if we move barely in from the coordinate the population will consist of one or two of allele 1 and a great many of allele 2. The 1 alleles will of course mate with 2 alleles (they are the only kind of mate they find) and will be present in heterozygous form. Hence K_{12} is the number of 2 alleles which will just prevent 1 alleles (in heterozygous form) from increasing. It is the carrying capacity of the environment for heterozygotes, expressed in units of 1 alleles. Similarly we can name the end points of curve $g(n_1,n_2) = 0$ as K_{22} and K_{21}. Then our graphs tell us that

(1) If $K_{11} > K_{21}$ and $K_{12} > K_{22}$ (i.e., the carrying capacity for 1 alleles is uniformly greater than for 2 alleles) as in Figure 52A, then the 1 alleles will oust the 2 alleles[1]. The opposite condition, in which the 2 alleles will oust the 1 alleles, is shown in part B of the same figure.

(2) If $K_{21} > K_{11}$ and $K_{12} > K_{22}$, (i.e., the carrying capacity for heterozygotes is greater than for both homozygotes) then both alleles will coexist indefinitely (part C of the figure).

(3) If $K_{21} < K_{11}$ and $K_{12} < K_{22}$ (carrying capacity of heterozygotes inferior to that of both homozygotes) then whichever allele is initially superabundant will oust the

[1] Unless the curves bend so strongly that they cross twice. In this case one intersection will be like curve C and the other like D.

other (part D). (This implies of course that a new mutant can never enter the population.)

We have now replaced the classical population genetics of expanding populations, where fitness was r, as measured in an uncrowded environment, by an analogous population genetics of crowded populations where fitness is K. For both situations we are assuming that no storms or other catastrophes interfere in such a way as to change the environment radically. As an example of how K selection and r selection can be in opposition, consider different situations in which crowding can either reduce the per capita food supply to a precariously low level, or else not have this effect. In an environment with no crowding (r *selection*), genotypes which harvest the most food (even if wastefully) will rear the largest families and be most fit. Evolution here favors *productivity*. At the other extreme, in a crowded area, (K *selection*), genotypes which can at least replace themselves with a small family at the lowest food level will win, the food density being lowered so that large families cannot be fed. Evolution here favors *efficiency* of conversion of food into offspring—there must be no waste. Margalef (1958) came to a similar conclusion by somewhat different reasoning.

The biogeographic implications of the distinction are considerable: where climates are rigorously seasonal and winter survivors recolonize each spring, in the presence of a bloom of foliage and food, we expect r selection favoring large productivity; where climates are uniformly benign, K selection and greater efficiency should result. This conclusion was anticipated in a less precise way by Dobzhansky (1950), who argued that competition and other forms of species interaction should influence evolution to a greater degree in the tropics than in the temperate zone.

Islands are interesting in this respect. A newly colonizing species will have r selection of course, as we argued in Chapter 4, but once it is safely established it will tend to shift back to K selection. In the common situation where the climate of islands is more uniform and moderate than in the nearby mainland, K selection should predominate relatively more than on the mainland. Hence species which are well and long established on an island, for example the endemics, should have evolved for efficiency while those which are perennially recolonizing (and going extinct) on small islands should evolve for productivity. Before we can test this prediction we must see what other forces will act on island species.

So far our account of selection has treated single species only. In other words we have only considered the implications of climatic and stability differences between islands and mainlands. But of course islands also differ in species diversity, and island species are in an environment with a greatly reduced number of species. The question which now might be raised is whether niches should be shifted or expanded to achieve great r or K when the number of competing species is reduced as on an island. In Chapter 5 we showed that the total harvest per unit time (hence the productivity of food) is greatest when habitat, but not food within habitat, is expanded in the absence of competition. Hence r selection should favor habitat expansion of island species. Efficiency is slightly more difficult to predict, being more closely related to the proportion of food items which are actually located during a food search. However, the island will not be crowded until all acceptable habitats and all suitable foods within each habitat are reduced by the dense population. Hence either in r selection or K selection it appears that habitat will expand on islands with a reduced number of competing species.

Clearly, also, r will be increased by a habitat or food shift which increases the density of available food. This will allow a temporary population increase—and suspension of K selection. When the environment is again crowded, the shift is a *fait accompli*.

These results apply to the abilities of single members of an island species: they should shift and expand in their habitat versatility. But the intrapopulation variability should also increase by means of various genetic devices such as polymorphism. At least this enrichment will always happen when the new genotype has a larger K or r in its chosen habitat than any previous occupant. With a few species the enrichment will cause niche expansion. With many it could even cause contraction, providing there were some correlation of habitat with feeding mechanism.

To add to the complexity of differences between island and mainland, the impoverishment of diversity on islands often results in an absence of effective predators. This is because the K of predators is considerably lower than that of their prey, so they are precariously rare even on relatively large islands. Hence on islands the compromise of selection will often lean less toward escape from predators. Consequently, island populations should be able to devote somewhat greater evolutionary attention to other virtues which conflict somewhat with effectiveness of escape.

So far these inferences on the distinctness of island selection rest on plausibility arguments and are in need of empirical confirmation. This is quite difficult because of the complexity of the factors involved. The only test of the theory of which we know is due to Cody (1966) who explained the clutch sizes of island birds in these terms. Cody used three coordinates representing evolutionary virtues between which some compromise is essential: large clutch, efficient food search, effective predator escape. He assumed that large clutch contributes to large r, that efficient food search contributes especially to large K, and that predator escape contributes to both, so that the absence of predators will not affect balance of clutch and feeding efficiency. Cody's argument then led to several correct predictions. On the seasonal temperate mainland where r selection is often more important, clutch size should be larger and feeding efficiency somewhat less. On the other hand, the effect should be reduced on offshore temperate islands, which enjoy a generally milder, less fluctuating climate. Toward the mainland tropics, where predators and K selection are supposed to be more important, the mainland compromise should lean considerably toward feeding efficiency and predator escape, while clutch size should suffer correspondingly. On tropical islands however, the reduction in clutch size should be considerably less—if any—because predators are not important and selection already leans toward feeding efficiency. Each of these inferences concurs with available data on geographic variation in clutch size; an example is given in Table 13. Thus modes of selection account for the geographic picture of clutch size variation; yet it would be premature to conclude, without much further documentation, that we have confirmed the model.

This discussion of the natural selection on islands will end with a bare mention of a controversial and little-understood subject—group selection. At its best, group selection is a poorly documented, uncertain phenomenon; and at its worst it is a pipe dream. Yet it has a possible mechanism—group extinction coupled with recolonization by random individuals. And on islands all the components of this mechanism are known to act. There are relatively frequent extinctions and recolonizations, and to the extent that the extinct populations perish because of poor genotype, group selection can be said to occur. Of course this is not to say that group selection has achieved any changes which would not have occurred under ordinary selection acting through survival of individuals or their kin. But writers who say that group selection is never to be mentioned if ordinary selection seems to offer a sufficient explanation are misusing "Occam's razor." The laws of natural selection are obeyed by group selection. The only new feature is that present fitnesses are not good predictors of future fitnesses; rather, the populations bring catastrophes on themselves in which fitnesses are drastically altered. In this sense even K selection is a form of group selection. However, it is a form in which the mechanism is clear.

The Stages of Post-colonization Evolution

We can now develop the picture of island evolution in more explicit form. Islands should be an excellent theater in which to study evolution. This is because a new insular population is set in one of the most favorable circumstances imaginable for rapid evolution. It has originated from a few individuals subject to the vagaries of random genetic

TABLE 13. Differences between clutch sizes (number of eggs per nest) of the same bird species on islands and nearby mainland localities of the same latitude. Clutch size averages smaller on the islands, as opposed to the mainland, in the temperate region but not in the tropical region. (From Cody, 1966.)

For temperate islands off the coast of New Zealand[a]			For tropical islands in the Caribbean[b]		
Species or genus	Average mainland clutch	Average island clutch	Species or genus	Average mainland clutch	Average island clutch
Anas spp.	8 (3 sp.)	3.5 (1 sp.)	Saltator albicollis	2.0	2.5
Bowdleria punctata	3.1	2.5	Tangara gyrola	2.0	2.0
Gerygone spp.	4.5 (1 sp.)	4.0 (1 sp.)	Habia rubica	2.3	2.0
Petroica macrocephala	3.5	3.0	Cacicus cela	2.0	2.0
Miro australis	2.6	2.5	Coereba flaveola	2.5	2.5
Anthornis melanura	3.5	3.0	Empidonax euleri	2.0	3.0
Cyanoramphus novaezelandeae	6.5	4.0	Elaenia flavogaster	2.3	2.5

[a] Species considered were either indigenous to the offshore islands and had a mainland relative in the same genus, or were subspecifically distinct on the island. The difference between the mainland and island means is 90% significant, by t-test.
[b] Species on isolated West Indian islands are compared to the mainland, and also the extreme southerly Lesser Antillean islands (isolated) are compared to Trinidad, which is very close to the mainland. The means are not significantly different.

sampling. It must pass through an early logarithmic growth phase in which a high reproductive potential is at a premium, and then out again to a new set of selective pressures; in other words, it shifts to some degree from K selection to r selection and then (quickly) back to K selection. It is isolated from the mother population. Finally, it is in a new environment that will ultimately direct its evolution away from that of the mother population.

We can conveniently think of the evolution of the new population as passing through three overlapping phases. First, the population is liable to respond to the effects of its initial small size. This change, if it occurs at all, will take place quickly—perhaps in only a few generations. The second phase, which can begin immediately and must continue indefinitely, is an adjustment to the novel features of the invaded environment. The third phase, an occasional outgrowth of the first two, consists of speciation, secondary emigration, and radiation.

The first phase: the founder effect. The evolutionary effects of initially small population size can only be guessed at this time. Conceivably two very different kinds of change can occur: (1) random genetic fixation and its aftermath due to natural selection and (2) adjustment of the intrinsic rate of increase by natural selection.

Mayr (1942, 1954, 1963) has explored the possibilities of random genetic effects in colonizing populations, discussing the phenomenon under the phrase "the founder principle." The founder principle is actually no more than the observation that a propagule should contain fewer genes than the entire mother population. The evolutionary hypothesis that Mayr based on the principle can be stated briefly as follows. Since the alleles of the founding population are relatively few in number, they must operate individually against a restricted, altered "genetic environment." And since the fitness of each allele is determined in part by the combination of genes in which it operates, in other words by its genetic environment, gene frequencies different from those of the mother population will be selected. Although genetic variability equivalent to that of the mother population will be gradually restored, the genetic composition will vary greatly. Mayr used this conjecture to explain the marked genetic differences often found among conspecific populations of islands which, superficially at least, appear to have similar environments. Dobzhansky and Pavlovsky (1957) conducted a laboratory model experiment which shows at least the feasibility of the random founder effect. When laboratory populations of *Drosophila pseudoobscura* containing equal proportions of two chromosome inversions were started with very small and moderately large groups of flies respectively, those from the very small groups displayed a greater between-group variance in equilibrial inversion frequencies than those from the large groups. This difference can be explained only as the outcome of differences in the total genic makeup among the groups due to sampling error. The smaller the sample size, the greater the expected variance among the samples. In fact, our theory of Chapter 4, which describes the probability of extinction of the descendants of a propagule, applies simultaneously to the genes in that propagule and can be used to make quantitative predictions of the founder effect.

On the other hand, feasibility does not at all imply effectiveness, a caveat that must be made doubly strong in all genetic drift theory. In fact, other theoretical considerations and some empirical evidence suggest that it may not be very important in the case of insular populations. On the theoretical side, islands that are colonized by a given species are probably colonized by multiple propagules. We have already seen that the impoverished state of a given taxon does not always mean that the island is so remote as to be reached only by rare propagules. This may be true in the extreme cases, such as the radiating taxa of the Hawaiian and Galápagos Islands, but in other cases impoverishment is more likely due to a combination of somewhat lower immigration rates and lower invasibility. Successful colonization can come about as easily by increased invasibility as by increased immigration rate. If the true immigration rate is high enough for as few as several sets of propagules to arrive per generation time, a large fraction of the total genetic variation of the mother population can be inserted into the founder population, in turn reducing random effects to a low level.

It is even less likely that genetic differences observed among real insular populations must be explained by sampling effects in default of selection hypotheses. We have already seen that all islands probably differ in biotic environment, even in the unlikely event that their physical environments are identical. Moreover, the smaller the islands the greater the variation among them. Since islands are also quite isolated in terms of gene exchange, their environmental differences alone should force genetic divergence. This is the conclusion reached by Ford and his co-workers (1964) in their studies of the butterfly *Maniola jurtina* in Great Britain. Theirs is one of the few adequate analyses of ecogenetic variation in insular populations to date. Differences in the relative frequencies of spotting patterns in the wings were found to be much more marked among populations on the small Scilly Islands than among those on the nearby English mainland. The adaptive significance of the spotting has not yet been deciphered, and the character states therefore fall in the large class of "nonsense" variation so familiar to taxonomists and ecologists. Yet the variation in this case is closely correlated with differences in prevailing habitats among the Scilly Islands, as well as temporal changes in habitat on at least one of the islands. Selection is clearly the major evolutionary force determining geographic variation in the character.

In summary, evolution due to genetic sampling error is an omnipresent possibility but one easily reduced to relative insignificance by small increases in propagule size, immigration rate, or selection pressure. In the current cycle of evolutionary writing, genetic drift is being deemphasized as a factor, principally because of a heightened appreciation of the magnitude and intricacy of natural selection revealed in recent field and laboratory studies. Although much doubt can also be cast on the special case of the founder effect, such an evaluation is still largely subjective and liable to change when direct measurements of the colonization process are finally made.

The first phase continued: changes in the life table. Lewontin (1965) has pointed out that not only will a high intrinsic rate of population increase preadapt species for successful colonization, but also the act of colonization itself will serve as a selection force favoring those genotypes that possess a higher r *within* the species. We would alter this (cf. Chapter 4) to r/λ. This r selection means that there will be a tendency for genotypes in very new insular populations to have a shorter developmental time, a longer reproductive life, and greater fecundity, in that order of probability (see pp. 84ff). The attractive feature of the hypothesis is that it should be readily testable, for example by comparing the life tables of individuals from populations subject to frequent extinction with those of individuals from conspecific populations in more stable environments. The conjectured selection should occur only briefly in the life of a given colonizing population, providing the population is allowed to grow to K and stay near there without unusual subsequent fluctuation. K will ordinarily be reached in very short time, in fact in only a few generations if the population has a good capability of surviving in the new environment. The mean r of the population should then subside back toward the mean r of the mother population. But very likely it will not equilibrate at exactly the same level, nor will the life-table parameters afterward be just the same, for the reason that evolution does not reverse itself perfectly—especially in a case where the environment is probably novel in various ways that can influence individual development. The degree of deflection and rebound of r depends of course on the heritability of the life-table parameters. Measurements of this heritability, and estimates of its influence on evolution under various colonizing conditions, remain to be made.

The second phase: adaptation to the new environment. Several characteristics of adaptation to insular environments can be generalized and documented. The most conspicuous is the tendency to lose dispersal power. The high frequency of flightlessness among endemic birds and insects on islands and mountain tops has been commented on for over a century. Darwin, in the *Origin of Species*, attempted to explain it as the outcome of selection opposing individuals that can take flight—and be wafted out to sea to perish. But Darlington (1943) pointed out that carabid beetle species on low, small islands tend to be winged, while those in sheltered montane habitats tend to be wingless. Both effects are directly opposite to what would be expected from Darwin's hypothesis. Darlington concluded that insects and birds lose the power of flight, and in some extreme cases insects go so far as to lose their wings, once long-distance dispersal is no longer needed. Populations on low, small islands are in constant peril of extinction due to storms, and a premium is therefore placed on genotypes able to disperse to new islands. But on mountains and larger, more rugged islands, long-distance dispersal is unnecessary, and genotypes are favored that do not invest energy in·the embryogenesis and maintenance of functional wings. Stated in the language of biogeographic theory, it is the mean dispersal distance of propagules (Λ, Chapter 6) and not the number of propagules that is reduced.

A parallel and equally graphic trend has been described by Carlquist (1966) in the Compositae of the Pacific islands. Endemic species of this plant family show a strong tendency to lose dispersal power through various combinations of several morphological changes: increase in fruit size, often as an adaptation to shady forest conditions, without concomitant increase in appendages that serve in dissemination; diminution or malformation of the appendages; and alteration of the mechanism of release of the fruits, such as the loss of the ability of the involucres to open at maturity.

A second means by which dispersal power is apt to be reduced is the tendency of evolving isolates to vacate the marginal habitats that are the best "staging" areas for departing and arriving propagules. In both ants and Compositae of the Pacific islands, evolution of local populations is accompanied by a penetration of the inner forest habitats and retreat from the less stable habitats located mostly around the coasts (Wilson, 1959; Carlquist, 1966).

A third, less-well-documented change that can further reduce dispersal power is a reduction in population size. Endemic species of insects, plants, and other groups of organisms on relatively species-rich islands often occur in restricted populations adapted to specialized local habitats. From the very fact that so many endemics tend to distribute in this pattern, it can be inferred that in some way restriction conveys greater stability and mean survival time to the population, while simultaneously reducing its colonizing power.

The relationship between the loss of dispersal power and speciation must be mutually reinforcing. In order to adapt to the local insular environment, species will on the average incur a loss of dispersal power. This loss in its turn tends to fragment the population into demes which specialize separately to the local environments in which they occur.

A second event which, in addition to loss of dispersal power, can be expected to occur in many populations is the shift in behavior or ecology which occurs as the species insinuates itself into its new set of competitors. Brown and Wilson (1956) showed that the pattern of evolved geographic variation expected to result from the shift, namely greater phenotypic difference between pairs of related species where they occur together than where they occur apart, is widespread in the insects and vertebrates. This "character displacement," however, has not occurred universally wherever the ranges of two similar species overlap. In many cases no divergence is known to occur, and in a few there is actually a convergence of the two species in some characters. The existence of displacement is therefore no more than a weak rule. Nevertheless, theoretical analysis of displacement, in conjunction with an examination of the better-analyzed examples, provides new insights into the role of species interactions in evolution.

Displacement can result from one or the other of two kinds of disoperation: hybridization, where the F_2 and subsequent hybrids are less fit than the parents, and competition.

159

Bossert (1963) has devised models to simulate the former case, concentrating on the population genetics of the various morphological, physiological, and behavioral traits that prevent species from exchanging genes. Among these traits, he distinguished *prezygotic mechanisms*, such as habitat and courtship pattern differences, which prevent fertilization, and *postzygotic mechanisms*, which contribute to lowered hybrid fitness when fertilization takes place. Displacement will occur only if (1) prezygotic mechanisms are imperfect and (2) the postzygotic mechanisms are strong enough to lower average hybrid fitness. As Mecham (1961) and others had recognized earlier, evolution is apt to occur most rapidly on the prezygotic side of the line, since the greatest advantage will accrue to the genotypes that can prevent fertilization and consequent gamete wastage. In his model Bossert assumed several simplifying conditions that can be relaxed extensively without damage to the qualitative results, for example an additive polygenic control of the prezygotic mechanism. Several of the results are of general interest to biogeographers. As exemplified in the sequence in Figure 53, the displacement can be expected to ensue rapidly, most of it being completed in only a few generations. If only one prezygotic mechanism is involved, such as a habitat preference or a courtship trait, the difference of the means of the two species will stabilize at about 30–50% of the common mean, the magnitude depending in part on the original variances of the two distributions.

The latter figure, 30–50%, is matched by most of the clear-cut cases of character displacement, an empirical generalization first made by Hutchinson (1959) and later reinforced by Schoener (1965). The curious thing is that the generalization holds for all examples of single-character displacement, including those which appear to involve competition rather than prezygotic barriers. This is in spite of the fact that competition cannot be readily conceptualized in the same way as prezygotic reinforcement, and there is no known theoretical reason why it should have the same exact quantitative effect on displacement.

Schoener (1965) noticed that interspecific differences in

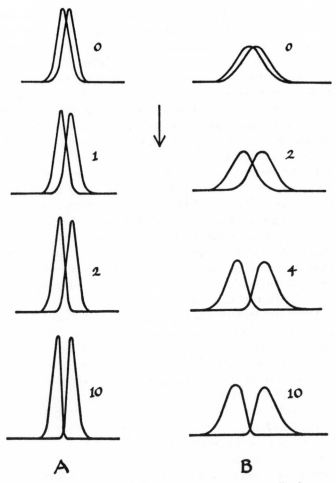

FIGURE 53. Predicted displacement between two equally dense species with zero hybrid fitness and only a single prezygotic mechanism. The prezygotic mechanism is a continuously varying character distributed as two overlapping frequency curves. Modified from Bossert (1963).

beak size in related bird species were greatest—around the 30–50% level—in three kinds of situations: where the food resource is sparsely distributed, as in birds of prey and kingfishers; where the species involved are large in comparison with other members of their taxonomic group; and on islands. Schoener reasoned that the species in all three situations have one feature in common—the inability to divide the environment efficiently among them according to habitat. As a result they divided it more according to food-particle size. The implication is that division of habitat is generally easier to achieve in all but the most restrictive circumstances. As Grant (1965) showed, many island birds have larger bills than their mainland counterparts, a fact which suggests a larger variance in food sizes. This is consistent with an unaltered range of acceptable items in the diet, however, for it could be caused by the island species' coming upon large and small items more often in the absence of competitors, but not accepting or rejecting them with altered probabilities.

There are two remaining results from Bossert's models that are of relevance to biogeographic theory. First, displacement will result initially in the two frequency curves skewing away from each other, and the degree of skewing is a measure of the rate of the displacement. As pointed out by Wilson (1965), the only data adequate to test the prediction are David Lack's (1947) measurements of displacing species of the *Geospiza fortis* group; these measurements are quite consistent with the prediction. The second result is a "conservation of momentum" theorem, which states that the rate of evolution of each of the two species is inversely proportional to the size of its population, that is,

$$\frac{d\Phi_1}{d\Phi_2} = \frac{n_2}{n_1},$$

where $d\Phi_1$ and $d\Phi_2$ are the rates of change in the displacing character in the two species, and n_1 and n_2 are the population sizes. Intuitively, the same result should hold for competing species as well as those reinforcing prezygotic barriers.

Our current knowledge of character displacement permits

a couple of new hypotheses about the minimal conditions for the coexistence of species or, put in the context of this chapter, the permissible level of interaction of jointly invading species. First, it appears that species must achieve a total difference, in the various ecological categories in which they compete, equivalent to a mean difference in one character of at least 30–50%. (The same condition applies, independently, to the prezygotic isolating mechanisms.) This is the equivalent of a total of 10^{-3} to 10^{-2} overlap in the joint distributions of the critical ecological characters (see Figure 54).

It is also apparent that if one of two interacting species is much scarcer than the other, and therefore contributes a very large share to the joint displacement, it is in danger of going extinct. The rare species will certainly go extinct if it is displaced faster than it can assemble the requisite characters by genetic recombination. Since most displacement will take place in only a few generations, this is nearly the equivalent of saying that the number of individuals outside the domain of the competitor (or hybridizing relative) must be as great at the very outset as the minimal number needed for survival. In other words, even genetic flexibility may do no good. Most species invade islands sequentially. As a result all the advantage goes to the species arriving first. It is conceivable that a given species which coexists on one island as the scarcer member of a pair can nevertheless exclude the commoner species from another, similar island if it has been lucky enough to arrive first. A second corollary is that where two species do invade sequentially, and any evolutionary displacement takes place, the later arrival will probably undergo the most change.

There is still another side to the displacement phenomenon. It was seen that in ants and composites, species tend to invade islands through ecologically marginal habitats. Later, as adaptation to the island progresses, they shift increasingly to central, species-rich habitats, such as the island forests. The correlation between weedy habits and dispersal ability demonstrated in other plant and animal groups (Baker and Stebbins, 1965) makes it probable that

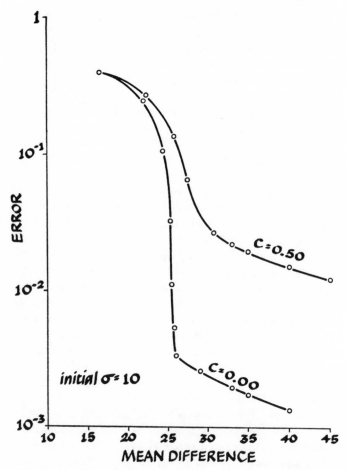

FIGURE 54. The degree of interaction of two species separated ecologically by a single character. The "error" is the frequency of hybrid matings in the case of an imperfect prezygotic isolating character. It can also be the percentage contacts of an interspecific nature in the case of an imperfect ecological difference separating two competing species and resulting in damage to both species. As the interspecific mean difference in the character is increased from about 15% of the common mean, the error drops rapidly until approximately the 30–50% level is reached. C is a measure of heritability; zero gives it a maximum value, 0.50 a very low value. (From Bossert, 1963.)

164

the process is a general one. It follows that if sequential invasions result in displacement, the direction taken by the older resident species will be toward the central habitats. Displacement will therefore hasten the speciation process in the older resident species, but not in the newcomer. Only by regaining the marginal habitats will the first resident recover its former dispersal capacity. Wilson (1959, 1961) has provided indirect evidence of several kinds to support the notion of such a "taxon cycle" in the Melanesian ant fauna; an outline of his hypothesis, together with some documentation of the marginal habitat phenomenon, is presented in Figures 55 and 56.

The reverse of character displacement can occur if a species invades a region, such as a smaller or more remote island, where there are fewer interacting species than in the source region. When Van Valen (1965) compared island and mainland populations belonging to six bird species

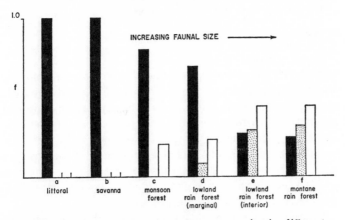

FIGURE 55. Proportions of ponerine ant species in different degrees of range expansion found in several major habitats on New Guinea. Dark columns: species widespread in Melanesia. Stippled columns: species restricted to single archipelagos but belonging to species groups centered in Asia or Australia. Blank columns: species restricted to single archipelagos and belonging to Melanesia-centered species groups. The marginal habitats, to the left, contain both smaller absolute numbers of species and higher percentages of widespread species. (From Wilson, 1959.)

165

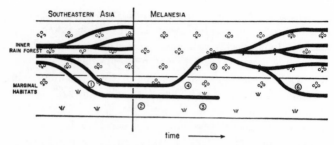

FIGURE 56. The inferred taxon cycle of ant species groups in Melanesia, in this case tracing the histories of groups derived ultimately from Asia. (1) Species or infraspecific populations adapt to marginal habitats in southeastern Asia, then cross the water gap to New Guinea and colonize marginal habitats there (2). In time these colonizing populations either become extinct (3) or invade the inner rain forests of New Guinea and surrounding islands (4). If they succeed in adapting to the inner rain forests, they eventually diverge to species level (5). As diversification thus progresses in Melanesia, the species in the group remaining in Asia may be contracting, so that in time the group as a whole becomes Melanesia-centered. A few of the Melanesian species, especially those on New Guinea, may re-adapt to the marginal habitats (6) and expand secondarily. (From Wilson, 1959.)

selected for convenience, he found that in five of the species the island populations occupied more habitats and displayed greater variability in bill size than the conspecific mainland populations. In fact, the variances in bill measurements in the island populations averaged twice those in the mainland populations. This kind of "character release" is the expected result if ecological release leads to destabilizing selection and the genetic fixation of phenotypic characters that permit the penetration of additional niches.

Character release might also be involved in the tendency for increase in size noted in island vertebrate populations by some authors. Grant (1965), in an example already mentioned, found that populations of birds on islands off the Pacific coast of North America and Mexico have both larger bills and longer tarsi than on the mainland. He ascribed these increases to the filling of food and perching niches of competitor species not present on the islands. In the islands of the Gulf of California, the size of individuals of the lizard *Uta stansburiana* is rather precisely related to the number of competitor species: the average size of individuals on a given island decreases as the square root of the number of coexisting species (Soulé, 1966). Other, less thoroughly analyzed cases of gigantism, as well as a variety of additional morphological trends in island biotas not yet covered by population theory, have recently been reviewed in the excellent popular book *Island Life* by Carlquist (1965).

Second phase: displacement versus convergence. We will now show that when competing species are brought together they need not always diverge from one another in evolution. Under certain conditions, which can be defined at least theoretically, they can even converge. Whether two species displace one another or converge depends both on their degree of specialization and on the productivity of the environment.

Earlier we introduced the concept of the grain of the environment. To eat an item of food, a member of a species must first come upon the item during its search, then it must take the item as food. The probability that a given item of food will be eaten by the predator is the product of the probability that the predator will come upon that item during its search, times the probability that it will eat the item after encountering it. The first of these terms (the probability of a successful search) is determined by the grain of the environment, defined relative to the species. If the environment is fine-grained, the species comes upon items of different resources in the proportion in which the resources occur. Hence if the populations of competing species are labeled as x and y and the quantities of two resources in fine-grained mixture are labeled as R_1 and R_2, we might find that

$$(7\text{-}1) \qquad \frac{dx}{dt} = x[i_1 R_1 + i_2 R_2 - T_x],$$

$$(7\text{-}2) \qquad \frac{dy}{dt} = y[j_1 R_1 + j_2 R_2 - T_y],$$

where i_1, i_2 and j_1, j_2 are the efficiencies with which the

167

resources are actually eaten, once they are discovered by the consuming species. In other words the i's and j's are determined by the second probabilities mentioned above, that once the items are found they will be eaten. The values T_x and T_y are the threshold quantities of the combined mixture of resources which allow x and y to maintain themselves. Since the resources are renewing, they are themselves governed by some equations, but these need not concern us at this point. From Equation (7-1), when $i_1R_1 + i_2R_2 > T_x$, then x will increase. We plot the inner boundary of this inequality as the solid line in the left graph in Figure 57,

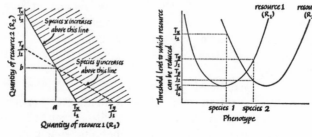

FIGURE 57. Two views of competition between two species for two resources, based on equations given in the text. The resources are defined as being renewable, divided into discrete particles, and mixed in a fine-grained manner relative to the species' search pattern.

in which R_1 and R_2 are the coordinates. Similarly, if $j_1R_1 + j_2R_2 > T_y$, then y will increase. We plot the inner boundary of this inequality as a dashed line in the same graph. Of course the dashed line will cross the solid line in the way shown only if $T_y/j_1 > T_x/i_1$ while $T_x/i_2 > T_y/j_2$, that is, if x can reduce R_1 to a lower level than y can, while y in turn is more effective in reducing R_2. If x were more effective in reducing both resources than y, then the dotted line would lie in the shaded region and x would keep the resources too sparse for y to persist. In the case depicted in Figure 57, x and y keep the resources R_1 and R_2 at the equilibrium level a, b. Any other species feeding on this same fine-grained mixture of R_1 and R_2 will have its own line on the graph, and it can invade the community consisting of x and y if its line lies inside the point a, b.

So far, this graphical model has provided a simple and not very illuminating picture of competition for discrete, renewing, fine-grained resources. We will now make it more useful by adapting it to the concepts of evolutionary convergence and displacement. To do this we must see whether mutant phenotypes have lines permitting "invasion" of the community, in other words whether new lines appear within the populations that fall inside a, b. Furthermore, we must see whether the successfully invading phenotypes of x and y are convergent or divergent. In the right-hand graph of Figure 57 a phenotype continuum is introduced, one-dimensional in the figure but conceivably either one-dimensional or multidimensional in nature. The ordinate in the graph is the value which was T_x/i_1 for our original species on resource 1, or T_y/j_1, and so forth. Clearly the right-hand graph, which gives us the end points of the line in the left-hand graph, contains the information on which phenotypes are competitively superior. To employ this theoretical information we now combine the graphs in two alternative cases, shown in Figure 58. Four phenotypes, two per species, are considered: x and y can be regarded as the original phenotypes of the two species, and x' and y' as the new phenotypes. First, examine the upper pair of graphs of Figure 58, in which the resource curves R_1 and R_2 are relatively close together. Since (in the right upper graph) the x' and y' lines intersect inside the point (a,b), x' and y' are competitively superior to x and y. Hence a *convergence* of species x and y would be favored.

In the second case (lower pair of graphs in Figure 58), x' and y' are inferior to x and y so that *divergence* is favored. The only difference between the two situations is that the R_1 and R_2 curves are shifted farther apart in this case. In other words R_1 and R_2 represent relatively similar resources in the convergence case and relatively different resources in the divergence case.

These purely theoretical results can be expressed in a more biological language, as follows. Species competing for

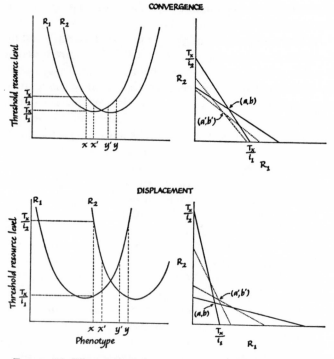

FIGURE 58. When new phenotypes (x' and y') are added to the pre-existing phenotypes (x and y) of the two competing species, evolution can occur. Whether the evolution will take the form of convergence or divergence depends on the shape of the resource-utilization curves and their distance apart. The geometric relationships of the curves determine in turn whether the equilibrium point (a',b') of the new phenotypes falls inside the old equilibrium point (a,b), resulting in their being competitively superior and causing convergence of the two species, or whether the opposite relation holds, causing displacement of the two species. A fuller explanation is given in the text.

a fine-grained mixture of similar resources will tend to converge to a jack-of-both-trades phenotype, while those competing for a fine-grained mixture of rather different resources will diverge, at least until they become specialists. The same result can be achieved by broadening and narrowing the R curves instead of shifting them together or apart. Consequently we can rephrase the results by saying that specialists, which generate narrow R curves, can coexist (without converging) on more similar resources than can generalists, which generate broad R curves.

The same results can be employed in still another way. First we note independently of the model that species should specialize under at least the two following conditions: (1) when the environment is productive, so that searching time for food items (as opposed to pursuit time) is low or (2) when species expend a relatively large proportion of their time and energy on pursuit, as in the case of carnivores pursuing very active prey. Expressing the situation loosely, we can say that when the pursuit/searching ratio is essentially zero, a species is only concerned with reducing the search time, which it does by enlarging its diet. In other words a searcher should generalize. When on the other hand the pursuit/searching ratio is very large, the pursuit time is relatively important. Pursuit time is reduced by specializing, and the pursuer should consequently specialize. Returning to the evolutionary model, it can be concluded that pursuers will tend to displace each other more than searchers. Also, a larger number of similar pursuer species should be able to coexist in a fine-grained environment than searchers. If this part of the theory has any value, such differences should be detectable in the relatively simple faunas of islands. Schoener's (1965) observation cited above (that birds of prey and kingfishers seem to rely on size rather than habitat differences to coexist) is also consistent with this theory. In fact, kingfishers and birds of prey are pursuers rather than searchers and hence can be expected to specialize and accordingly to coexist by means of size differences alone. As Schoener showed, many other species do not seem able to coexist except by habitat separation.

Species packing in the second phase. We now come to the intriguing question, By how much can adaptation to the environment increase species numbers on the island? Suppose we begin with an island that is being filled from a nearby species pool much larger in size than its own equilibrial number. Providing immigration rates are not excessively low, a short period of time should see a buildup to a level close to the maximum potential equilibrial number that can be achieved with any combination of species from the species pool. If the species pool is very large, and the source region at all ecologically different from the recipient island, a longer period of time will permit a steady evolutionary adjustment of the colonist species to each other and to the new environment and, as a consequence, a steady increase in the species number. Viewed, then, in terms of ecological time spanning a few generations, the original number of species is at or close to equilibrium. But seen in evolutionary time spanning hundreds or thousands of generations, the original number is to be regarded only at quasi-equilibrium: it is destined to creep steadily upward. The upper limit of the quasi-equilibrium will presumably be set in most cases by limitations in the climatic and geographical stability of the region.

Recently Wilson and Taylor (1967) exploited an unusual set of circumstances to make an empirical estimate of the potential increase in species densities in the Pacific ant fauna. Few if any ant species are native to Polynesia east of the line connecting Samoa and Tonga. The central and eastern archipelagos are populated in large part by 35 "tramp" species carried there from various parts of the tropics by human commerce. No one island contains all of the tramp species, and most contain less than one-fourth. Several lines of evidence suggest that the species densities have nevertheless stabilized. Samoa, however, contains an older, native fauna drawn from a pool estimated to contain about 43 species of Indo-Australian origin. The tramp and native species pools are therefore comparable in size. By comparing the number of native species on Upolu with the expected number of tramp species on an island of comparable size, an estimate of the potential increase in density of species of tramp origin could then be made. Under the conditions that (1) the immigration rates hold constant and (2) the tramp faunas persist as long as the native Samoan fauna, the potential factor of increase was estimated to lie between $1.5\times$ and $2\times$. And since the slopes of the area-diversity curves do not vary much among faunas, the same

172

factor of increase should hold for islands of all areas (Figure 59).

The third phase: speciation and radiation. Given enough time, all insular populations will evolve away from one another and from the mother population. Eventually they must reach a level of divergence sufficient to be judged subjectively as subspecies and, finally, as species. Certainly not all native populations have achieved either status; many are indistinguishable from other populations of the same

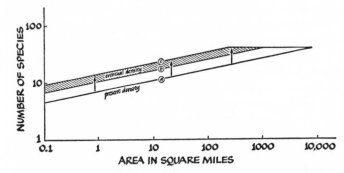

FIGURE 59. The estimated potential evolutionary increase in species density of the Polynesian ant fauna, a newly assembled set of species of diverse origins. (a) the present curve based on the relatively high values from Upolu and Fakaofo. (b) and (c) increase of 1.5× and 2× respectively, the estimated limits.

species. The reason cannot be that they are insufficiently isolated. Their degree of isolation is immense compared with that of contiguous but genetically differentiated populations on the mainlands. The most obvious explanation is that they are still relatively youthful. But youthfulness in members of an equilibrial fauna implies extinction of other, former members. In fact, the percentage of non-endemic species is probably a measure of the turnover rate. This inference can be tested. We know on both theoretical and empirical grounds that the turnover rate of species varies inversely with island area. It follows that percentage endemicity should increase with island area. Data given by Wilson (1961) for insular ant faunas and by Mayr (1965b) for insular bird faunas are consistent with the prediction (see Figures 40 and 60).

Once a population has evolved to full species level, the stage is set for adaptive radiation. On islands holding equi-

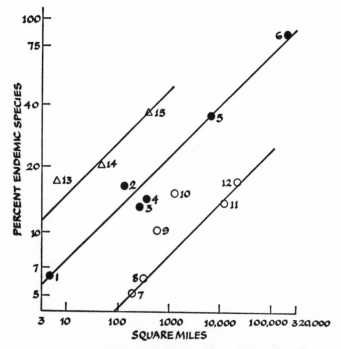

FIGURE 60. Percentage of resident bird species that are endemics as a function of island area, in three kinds of islands. Dark circles: Solitary, well-isolated islands: Lord Howe, 1; Ponape, 2; Rennell, 3; Chatham, 4; New Caledonia, 5; Madagascar, 6. Open circles: Single islands near mainlands or large archipelagos: St. Matthias, 7; Pemba, 8; Manus, 9; Socotra, 10; Timor, 11; Tasmania, 12. Triangles: Islands in the Gulf of Guinea: Annobon, 13; Principe, 14; San Tomé, 15. (Modified from Mayr, 1965b.)

librial biotas, the ratio of the number of species arriving from other islands in the same archipelago to the number arriving from outside the archipelago can be expected to increase with distance from the major extra-archipelagic source area. Where the archipelagos are of approximately similar area and configuration, the ratio should increase in an orderly fashion with distance. The accumulation on single islands of species generated within the archipelago provides the best available measure of what is loosely referred to in the literature as adaptive radiation. Specifically, adaptive radiation takes place as species, when generated within archipelagos, disperse between islands, and, most importantly, accumulate on individual islands to form diversified associations of sympatric species. In equilibrial biotas, then, the following prediction is possible: adaptive radiation will increase with distance from the major source region and, after corrections for area and climate, reach a maximum on archipelagos and large islands located in a circular zone close to the outermost dispersal range of the taxon. This has been referred to as the "radiation zone" of taxa with equilibrial biotas (MacArthur and Wilson, 1963). Many examples possibly conforming to such a rule can be cited: the birds of Hawaii and the Galápagos, the murid rodents of Luzon, the cyprinid fish of Mindanao, the frogs of the Seychelles, the gekkonid lizards of New Caledonia, the Drosophilidae of Hawaii, the ants of Fiji and New Caledonia, and others (see especially in Darlington, 1957; and Zimmerman, 1948). But there are conspicuous exceptions: the frogs just reach New Zealand but have not radiated there; the same is true of the insectivores of the Greater Antilles, the terrestrial mammals of the Solomons, the snakes of Fiji, and the lizards of Fiji and Samoa. To say that the latter taxa have only recently reached the islands in question, or that they are not in equilibrium for some other reason, would be a premature if not facile explanation. But it is worth considering as a working hypothesis.

Saturation, Impoverishment, and Harmony

These three terms, seemingly so pregnant with meaning, are used frequently in the biogeographic literature. The ecological concepts to which they allude are important but ill formed, so that an explicit restatement of their meaning in terms of the new theory is needed.

Saturation implies equilibrium. In a saturated (equilibrial) biota the successful colonization of new species means that an approximately equal number of older, resident species must become extinct. We would define saturation, then, simply as the equilibrial state. As shown by Wilson and Taylor (1967), an equilibrium can be defined in either ecological or evolutionary time. In ecological time a balance is struck between immigration and extinction of species during a period of time too brief for significant evolutionary change to occur. This is saturation in the ordinary sense of the word. In evolutionary time, a gradual increase in the species number is permitted through adaptation of the immigrant species to each other and to the local environment over a relatively long period of time. It is also conceivable that an alteration could come about through evolutionary changes in potential immigrant species in the source region. Viewed through evolutionary time, the equilibrium seen in the life-time of a human observer is really a quasi-equilibrium.

An impoverished or depauperate biota is one in which the species density is relatively very low. Impoverishment implies nothing about the degree of saturation. It can characterize the biota of a newly formed island in the early stages of colonization, at which time the species number is well below equilibrium; it can equally well characterize islands whose species are at equilibrium but which are so isolated or ecologically poor as to have low equilibrial numbers.

By increasing the immigration rate, an impoverished biota can be changed into a richer one, yet without altering its equilibrial condition in the end. It will merely shift from one saturated state to another. Consider for example the land and fresh-water birds of Hawaii. Before the advent of European settlement, there were only about 40 species. Yet, as we have argued in Chapter 3, the fauna was probably saturated. In other words, over long intervals of time the extinction rate equaled the immigration rate; for every new species that successfully colonized the archipelago, an average of one species became extinct. Perhaps there were very slow secular changes in the total number of species owing to the ever-improving adaptation of the fauna as a whole to the Hawaiian environment, but the resulting overall increment in new species per unit time must have been insignificant in comparison to the constant turnover rate. With the settlement of Hawaii in the nineteenth century, however, and the deliberate introduction of new bird species into the islands by "Acclimatization Societies," the immigration rate was greatly increased. This had the effect of shifting the immigration curve from the equivalent of "far" in the direction of "near" in Figure 8. If the new inpouring of immigrants were to be held constant, the number of Hawaiian bird species—native plus introduced—would move to a new, much higher equilibrium level. In this special sense, therefore, the old Hawaiian fauna can be regarded as "impoverished." If, on the other hand, all further importations were strictly forbidden so that the immigration rate returned to the old, natural level, the number of species might gradually decline to a third equilibrium not radically different from the pre-European level. In short, the temporary injection of new immigrants does not ensure the permanent enrichment of an insular fauna. Only a permanently increased immigration rate can do that.

A harmonic (or "balanced") biota is one that contains all of the basic adaptive types found in other ecologically comparable regions. This condition will be approached by an island which is close enough to a rich source region to receive most of the species groups among its immigrants and which is large and diversified enough to sustain them. It can also be approached by an isolated continent or archipelago if the land mass is large enough and stable enough to permit the few stocks reaching it to undergo full adaptive radiation. Thus Australia and South America acquired harmonic mammal faunas independent of those in the rest of the world, but Madagascar and the West Indies did not. An isolated island or archipelago can be both harmonic and saturated with respect to one taxon but disharmonic and unsaturated with respect to a superordinate taxon. In Hawaii, certain insect families, such as the Gryllidae, Coenagriidae, Lygaeidae, Miridae, Delphacidae, Hemerobiidae, Cosmopterygidae, Carabidae, Anobiidae, Drosophilidae, etc., are represented by diverse assemblages of species derived from few stocks (Zimmerman, 1948). In some cases they have radiated to an extent exceeding that of the family in the entire remainder of the world, and consequently by definition are harmonic. But the orders to which they belong are notably disharmonic. Many families in the same orders which are extremely diverse and ecologically dominant in other parts of the world, such as the Scarabaeidae and Formicidae, are entirely absent. No less than 21 orders of insects are also absent. The missing insect groups are usually either poor dispersers or else their preferred habitat is wholly missing. In the former case the islands are clearly "unfilled," and recent introductions by man have met with spectacular success. In the past two hundred years, for example, the Hawaiian islands have acquired 36 species of ants (Formicidae) belonging to 21 genera, which have become ecological dominants over virtually the entire land area.

SUMMARY

In defining the peculiarities of post-colonization evolution, a fundamental distinction was made between r selection and K selection. The intrinsic rate of population increase, r, is likely to be increased in the earliest stages of colonization, when population growth is unrestricted. Moreover, r will be held at a high value by those species whose histories include frequently repeated colonizing episodes. But most species occupying stable habitats, once they have attained their maximum population size, K, will tend once again to reduce r. There will be a simultaneous tendency to increase K through finer adaptation to the local environment. Thereafter, the relative amounts of r selection and K selection will be determined by the stability of the local environment.

Changes in r and K imply changes in life-history parameters and efficiency of resource utilization respectively. As the study of these subjects progresses (Chapters 4 and 5), it should be possible to make increasingly sound predictions concerning the ecology of colonizing species at different stages of the colonizing episode.

Founding populations are typically very small in size and contain only a small part of the genetic variation of the mother population. Insofar as the genes received are determined by chance, there will be an indeterminate component in the genetic variance *among* founding populations. The actual contribution of this "founder effect" to evolution can be assessed only by empirical field studies. To date such studies have been too few to permit any broad conclusions.

As the first evolutionary phase runs its course, during which r may be altered and the founder effect can occur, the population begins a long-range adaptation to the peculiarities of the local environment. Few generalizations can be made about this second, even more important phase. Dispersal power is commonly reduced, often to a striking degree. In addition, adjustments are frequently made to newly acquired competitor species. Under certain circumstances, which can be predicted by resource-utilization models, the species should increase the difference between it and its competitors—providing it is able to survive at all. Under other conditions, the competing species should converge. Another form of displacement occurs when characters used as prezygotic isolating mechanisms are imperfect and permit the production of less fit hybrids. In that case selection will tend to favor the enhancement of differences in the isolating characters. Evolutionary displacement can be expected to show certain general features, such as stabilization of the affected characters at about a 0.001 to 0.01 overlap in the joint distributions, a skewing of the affected frequency curves of the species away from each other, and a relative amount of displacement inversely proportional to the relative abundance of the species.

When a species colonizes an island that happens to be free of its former competitors, evolutionary release can (and often does) occur. There is a tendency for the variances of the ecologically important characters to increase and for the species to converge in the direction of one or more of the absent competitors. Displacement and release can occur in a complex, alternating fashion as species expand and contract their ranges, or as expanding species penetrate species-rich and species-poor archipelagos in alternation.

Evolution on islands and archipelagos can eventually lead to the formation of new, autochthonous species. In order for evolution to proceed to this degree, islands must be relatively large and stable, otherwise populations will not survive long enough to undergo sufficient local adaptation. Near the outer limit of the dispersal range of a given taxon, speciation and exchange of newly formed autochthonous species within an archipelago can outrun immigration from outside the archipelago and lead to the accumulation of species on single islands. Despite their common origin, such species tend to be adaptively quite different from each other, and the result is adaptive radiation in the strict sense.

CHAPTER EIGHT

Prospect

Biogeography has long remained in a natural history phase, accumulating information about the distribution of species and higher taxa and the taxonomic composition of biotas. Interpretive reasoning has been largely directed to the solution of special problems connected with the histories of individual taxa and biotas. Without doubt this descriptive activity will continue to be of fundamental importance to the science, one of the most physically adventurous of all scientific enterprises and, in the richness of the detail it unfolds, esthetically pleasing. But biogeography is also in a position to enter an equally interesting experimental and theoretical phase.

There is no reason why the biotas of islands cannot be manipulated experimentally. There are many places in the world where islands are both very small and numerous: for example, the red mangrove islets of many tropical countries, the sand islets of the Caribbean, the Indian Ocean, the forested islets of the Canadian lakes, the lakes themselves, the hardwood "hammocks" of tropical grassland, the coniferous enclaves in tundra and on mountain tops. In these little places it is possible to remove elements of a biota or the entire biota, manually or by poisoning, or to add elements. Miniature "Krakataus" can be generated at will and in sufficient replication to yield statistically sound results. In this book we have provided theoretical arguments, together with some documentation, which indicate that on very small islands the process of natural extinction is accelerated. Immigration, colonization, and turnover should be measurable in a reasonably short time span, certainly covering no more than a few years and perhaps, in the case of invertebrate animals and annual plants, months or less. By careful manipulation the processes can be speeded, slowed, and directed in a way to yield new information about them. The smallness and abundance of the islands moreover ensure that experiments can be conducted without significantly altering the environment of the surrounding areas. Their own biotas can be reconstituted at the end of the experiments. Where microorganisms are concerned, the "islands" can be created artificially, as Bassett Maguire has already done with bottles of sterile water and Ruth Patrick with glass slides. Such synthetic studies can reveal much of general significance but of course nothing about peculiarities in the dispersal and colonizing behavior of the higher plants and animals which comprise most of the several millions of living species on earth.

Certainly a meshing of new experimental information with quantitative theory could galvanize biogeography and have extensive repercussions in ecology and evolutionary theory. What, then, are the prospects for biogeographic theory? By confining ourselves to islands in this brief book we have intentionally eliminated many of the most troublesome— and interesting—problems. A truly comprehensive theory of biogeography will treat islands and continents together. It will recognize that island populations are fragmented and the boundaries of their subunits set by factors having little to do with innate biological properties. Hence future theory will concentrate on the boundaries of species ranges as they are encountered on ecologically uniform or continuously varying terrain. By the same token, account will have to be taken of the geometric structure of populations. Population geometry must be related in important ways to the adaptive strategy of species and, through that, to the colonization behavior and the multiplication of species. Existing speciation theory, which is also still in a rudimentary state, provides us at most with slight clues.

Global patterns of distribution also need to be reconsidered. We know that species diversity, relative abundance, and population geometry change with climate. Such variation affects on the one hand the structure, stability, and energy flow of the plant and animal communities. It must also affect the rate and perhaps mode of speciation, together with the propensity for adaptive radiation. Almost certainly it determines the dominance of the biota of one region over that of another, the global patterns of which have been already forcefully documented by Matthew (1939), Simpson (1950) and Darlington (1957).

In short, biogeography appears to us to have developed to the extent that it can be reformulated in terms of the first principles of population ecology and genetics. In order to achieve this restatement, it might be necessary to de-emphasize for the moment the traditional problems concerning the distribution of higher taxa and the role of geological change in the determination of this distribution and to turn instead to detailed studies of selected species. A "biogeography of the species" requires both theory and experiments that must be in large part novel. Simultaneously it demands a cultivation of population and community ecology in a way that contains much more evolutionary interpretation than has been traditional.

Glossary

Adaptive radiation. The multiplication of species (speciation) which occurs under the special circumstances that lead to the coexistence on single islands (or continents) of closely related but adaptively very different species.

Area-diversity curve. The curve relating numbers of species (or some other measure of species diversity) to areas of the different islands. When numbers of species are used, the curve is referred to as the area-species curve; this special relation approximately fits the equation $S = CA^z$, where S is the number of species, A is the area, and C and z are fitted constants. Usually the area-species curve is presented in a log-log plot, so that z can be readily approximated as the slope of the curve.

Area-species curve. See area-diversity curve.

Autochthonous. Referring to a species or some other taxon that originated in the region stipulated. The taxon itself is sometimes called an autochthon of the region.

Biogeography. The study of the distribution of species of organisms over the face of the earth. Biogeography is concerned with the limits and geometric structure of individual species populations and with the differences in biotas at various points on the earth's surface. The local, ecological distribution of species, together with such synecological features as the structure of the food web, are treated under biogeography only insofar as they relate to the broader aspects of distribution.

Biota. The plants (flora), animals (fauna), and microorganisms (flora and fauna) all taken together.

Character displacement. The process of genetic divergence of two (or more) species when they come together, due to the harmful interaction of certain members of one species with individuals of the other species. The harmful interaction consists either of hybridization resulting in offspring with reduced fitness or else of interspecific competition. As a consequence those phenotypes that most closely resemble phenotypes in the opposing species, with reference to the critical characters that affect mating and competition, are differentially eliminated, and the species diverge. Displacement is most readily detected as an increased phenotypic difference between the species where their ranges overlap contrasted with a less marked difference where the species occur apart.

Character release. The opposite of character displacement (*q.v.*): an increase in variance in one or more phenotypic characters due to the removal of a hybridizing or competing species. (See also *ecological release.*)

Coarse-grained. Pertaining to two or more resources distributed in such a way that the consumer species encounters them in a proportion different from that in which they actually occur. Opposed to fine-grained (*q.v.*).

Colonization. The relatively lengthy persistence of an immigrant species on an island, especially where breeding and population increase are accomplished.

Colonization curve. The change through time of numbers of species found together on an island.

Compression. The decrease in niche width due to interaction with competing species.

Compression hypothesis. From *a priori* considerations, species faced with competition should decrease the number of habitats in which they live rather than the number of kinds of food items they eat within a given habitat.

Conspecific. Belonging to the same species.

Density-dependent. Pertaining to any numerical property, such as birth rate, death rate, or species extinction rate, which changes with change in density of organisms (or of species).

Depletion effect. The decrease in immigration rate of species onto an island with an increase of species resident on the island, in other words, a descending immigration curve.

Disharmonic. Referring to a biota containing only a small proportion of the basic adaptive types found in surrounding source regions. Some previous writers have used the term to apply to a biota with a small proportion of the species stocks found in surrounding regions, which in turn implies a restriction in the array of basic adaptive types. But by explicitly employing the idea of adaptive types, we call attention to the more interesting ecological phenomena involved.

Displacement. See *character displacement.*

Diversity. Either the absolute number of species present, or some measure that incorporates both the number of species and their relative abundance, e.g., $\sum_i p_i \ln p_i$, where p_i is the relative abundance of the ith species; or $1 / \sum_i p_i^2$.

Dry Tortugas. The outermost of the Florida Keys, which are small islands located just south of the Florida peninsula. The Dry Tortugas consist of seven small islands located sixty miles west of Key West.

Ecological release. The enlargement of the niche of a species due to the removal of a competitor or some other species whose presence would be restricting. The enlargement may or may not be accompanied by an increase in the variance of certain phenotypic characters, which increase is referred to as *character release* (*q.v.*).

Endemic. A species or other taxon is said to be endemic to a particular region if it is native only to that region. The taxon itself is often referred to as "an endemic" of the region.

Equilibrium. The state in which rate of death of organisms in a population of organisms equals rate of birth, or rate of extinction of species in a biota equals rate of immigration of new species.

Eurytopic. Found in many different habitats.

Exponential dispersal. Dispersal in which the number of surviving propagules falls off with increasing distance according to an exponential density distribution.

Extinction. The total disappearance of a species from an island (does not preclude recolonization).

Extinction curve. The curve relating the extinction rate to the number of species present, or to time; in the latter case it is called the time curve of the extinction rate.

Extinction rate. Number of species on an island that become extinct per unit time.

Fauna. The animal species of a particular region. The term also refers to the species of a given taxon, such as the birds or cicindelid beetles, found in the region. A fauna can also be somewhat more precisely defined as a set of species in a region isolated enough so that the set is more or less peculiar to it, as opposed to faunula (*q.v.*)

Faunula. A set of animal species found in a relatively small, poorly isolated region and not peculiar to it.

Fine-grained. Pertaining to two or more resources distributed in such a way that the consumer species encounters them in the same proportion in which they actually occur. Opposed to *coarse-grained* (*q.v.*).

Flora. The plant species of a particular region.

Founder effect. A genetic alteration in a colonizing population that is ultimately due to chance deviations in the makeup or proportions of the genes in the propagules from those in the source population or other colonizing populations. The divergence can be due to sampling error alone, i.e., genetic drift, or to an interaction between sampling error and selection. (See also *founder principle.*)

Founder principle. The principle that propagules starting a new population will contain fewer genes than the mother population from which they originated. The founder principle can, theoretically at least, lead to the founder effect.

Harmonic biota. As loosely defined in Chapter 7, a biota that contains the basic adaptive types found in other, ecologically comparable regions. (See *disharmonic.*)

Immigration. The process of arrival of a propagule (*q.v.*) on an island not occupied by the species. The fact of an immigration implies nothing concerning the subsequent duration of the propagule or of its descendants.

Immigration curve. The curve relating the immigration rate to number of species already present, or to time; in the latter case it is referred to as the time curve of the immigration rate.

Immigration rate. Number of new species arriving on an island per unit time.

Impoverishment. Process leading to a relatively low number of species or a low value of any other measure of diversity.

Individuals Curve. The curve of the numbers of individual organisms found in each species-abundance class. The number of organisms found in the species-abundance classes is given on the abscissa, and the number of species belonging to the species-abundance classes is given on the ordinate.

Invasibility. The ease with which a community of species can be invaded by a new immigrant species. There can occur a packing so tight that no invasion is possible.

Isocline. A line along which some variable retains a constant value. The method of isoclines is an important graphical technique for solving simultaneous non-linear equations.

K selection. Selection favoring a more efficient utilization of resources, such as a closer cropping of the food supply. This form of selection will be more pronounced when the species is at or near K. Opposed to *r selection* (*q.v.*).

Key. A small island.

Krakatau. One of the Krakatau Islands, located in the Sunda Strait between Java and Sumatra.

Local species. A species limited to a relatively small area; in terms of ordinary collector's data, it is a species that has been found in only one or a very few localities.

Lognormal distribution. A distribution which is normal when the measure of quantity (along the abscissa) but not the frequency (along the ordinate) is arrayed on a logarithmic scale. Many species-abundance curves can be approximated by the lognormal distribution.

Marginal habitats. Habitats containing relatively low species diversity. The impoverishment is sometimes due to marginal physical conditions and sometimes to other causes.

Mean survival time. The average life span of a population. See T_s in list of symbols.

Normal dispersal. Dispersal in which the numbers of surviving propagules falls off with increasing distance according to a normal density distribution.

Postzygotic. Pertaining to intrinsic (genetically determined) characters that reduce fitness of hybrid offspring.

Prezygotic. Pertaining to intrinsic (genetically determined) characters that inhibit mating between populations or successful fertilization if mating does occur.

Propagule. The minimal number of individuals of a species capable of successfully colonizing a habitable island. A single mated female, an adult female and a male, or a whole social group may be propagules, providing they are the minimal unit required.

Quasi-equilibrium. An apparent equilibrium in which species density is nonetheless increasing through evolution; thus if measurements were taken over long periods of time, the immigration rate would be seen to slightly exceed the extinction rate.

r selection. Selection favoring a higher population growth rate and higher productivity. This form of selection will come to the fore during the colonizing episode, or in species which are frequently engaged in colonizing episodes and hence must frequently build back up to K. Opposed to K *selection (q.v.).*

Radiation. See *adaptive radiation.*

Radiation zone. A zone, near the outer limits of the distribution of a taxon, where immigration from outside archipelagos is so rare that speciation and radiation occur easily within the archipelago.

Reproductive value. A measure, labeled by the symbol v_x, of the expected number of offspring yet to be produced by an individual of a given age. In biogeographic terms, the value v_x of an x-year-old may be defined as the expected number of individuals in a colony (at some remote future time) founded by a propagule of x-year-olds.

Saturation. The equilibrial condition, i.e., the state at which immigration is balanced by extinction.

Speciation. As narrowly defined here, the multiplication of species, i.e., the evolution of two or more contemporary species from one ancestral species.

Species-abundance curve. The frequency curve of species containing various numbers of individual organisms in their populations, i.e., a plot of species numbers (given on the ordinate) against abundance of organisms per species (given on the abscissa).

Species curve. The species-abundance curve. The term used by Preston (1962); see Chapter 2.

Species packing. The process of increasing the numbers of species, which can involve ecological changes in the resident species or immigrant species or both. There is a limit to the closeness of packing.

Species pool. The number of species able to immigrate to an island, whether they can persist there or not.

Staging area. A habitat or particular locality from which, by virtue of its structure or location, propagules are unusually likely to cross barriers or proceed outward for long distances.

Staging habitat. See *staging area.*

Stenotopic. Found in only one or a relatively small number of habitats.

Stepping stone. An island used by a species in spreading from one region to another.

Taxon. A set of species: either a single species or a group of species which together constitute a higher taxonomic unit such as a genus or family.

Taxon cycle. In the Pacific ant fauna, the inferred cyclical evolution of species, from the ability to live in marginal habitats and disperse widely, to preference for more central, species-rich habitats with an associated loss of dispersal ability, and back again.

Turnover. The process of extinction of some species and their replacement by other species.

Turnover rate. The number of species eliminated and replaced per unit time.

References

Allan, H. H., 1961, *Flora of New Zealand*, Vol. 1. R. E. Owen, Government Printer, Wellington, New Zealand.

Baker, H. G. and G. L. Stebbins, 1965, *The Genetics of Colonizing Species*. Academic Press.

Blake, E. R., 1953, *Birds of Mexico*. University of Chicago Press.

Bossert, W. H., 1963, Simulation of character displacement in animals. Ph.D. Thesis, Dept. of Applied Mathematics, Harvard University,

Bowman, H. H. M., 1918, Botanical ecology of the Dry Tortugas. *Carnegie Inst. Wash. Papers Dept. Marine Biol.*, 12: 111–138.

Brown, T., 1893, *Manual of New Zealand Coleoptera*. New Zealand Geological Survey.

Brown, W. L. and E. O. Wilson, 1956, Character displacement. *Syst. Zool.*, 5: 49–64.

Cameron, W. A., 1958. Mammals of the islands in the Gulf of St. Lawrence. *Natl. Museum Canada, Bull.* 154.

Carlquist, S., 1965, *Island Life: A Natural History of the Islands of the World*. Natural History Press, Garden City, New York.

——, 1966, The biota of long-distance dispersal. II. Loss of dispersability in Pacific Compositae. *Evolution*, 20: 30–48.

Cockbain, A. J., 1961, Fuel utilization and duration of tethered flight in *Aphis fabae* Scop. *J. Exptl. Biol.*, 38: 163–174.

Cody, M. L., 1966, A general theory of clutch size. *Evolution*, 20: 174–184.

Cole, L. C., 1954, The population consequences of life history phenomena. *Quart. Rev. Biol.*, 29: 103–137.

Cooke, C. M., Jr., and Y. Kondo, 1960, Revision of Tornatellinidae and Achatinellidae (Gastropoda, Pulmonata). *Bull. B. P. Bishop Museum*, 221: 1–303.

Crowell, K., 1961, The effects of reduced competition in birds. *Proc. Natl. Acad Sci. U.S.*, 47: 240–243.

Curtis, J. T., 1956, The modification of mid-latitude grasslands and forests by man. In W. L. Thomas, Jr., ed., *Man's Role in Changing the Face of the Earth*. University of Chicago Press.

Dammerman, K. W., 1948, The fauna of Krakatau 1883–1933. *Verhandel. Koninkl. Ned. Akad. Wetenschap. Afdel. Natuurk.*, (2)44: 1–594.

Darlington, P. J., 1943, Carabidae of mountains and islands: data on the evolution of isolated faunas, and on atrophy of wings. *Ecol. Monographs*, 13: 37–61.

——, 1957, *Zoogeography: the Geographical Distribution of Animals*. Wiley.

Darwin, C. R., 1963, Darwin's ornithological notes. (Edited with introduction, notes, and appendix by Nora Barlow.) *Bull. Brit. Museum* (Nat. Hist.), *Hist. Ser.*, 2: 203–278.

Davis, J. H., 1942, The ecology of the vegetation and topography of the Sand Keys of Florida. *Carnegie Inst. Wash. Publ.*, 524: 113–195.

Dobzhansky, T., 1950, Evolution in the tropics. *Am. Scientist*, 38: 209–221.

Dobzhansky, T. and O. Pavlovsky, 1957, An experimental study of interaction between genetic drift and natural selection. *Evolution*, 11: 311–319.

Docters van Leeuwen, W. M., 1936, Krakatau, 1883 to 1933. *Ann. Jard. Botan. Buitenzorg*, 56–57: 1–506.

Duellman, W. E. and A. Schwartz, 1958, Amphibians and reptiles of southern Florida. *Bull. Florida State Museum*, 3: 181–324.

Ehrendorfer, F., 1965, Dispersal mechanisms, genetic systems, and colonizing abilities in some flowering plant families. Pp. 331–352 in H. G. Baker and G. L. Stebbins, eds., *The Genetics of Colonizing Species*. Academic Press.

Eisenmann, E., 1952, Annotated list of birds of Barro Colorado Island, Panama Canal Zone. *Smithsonian Inst. Misc. Collections*, Vol. 117, No. 5.

Feller, W., 1957, *An Introduction to Probability Theory and Its Applications*, Vol. 1, 2nd ed. Wiley.

Fisher, R. A., 1930, *The Genetical Theory of Natural Selection*. Oxford University Press.

Ford, E. B., 1964, *Ecological Genetics*. Wiley.

French, R. A., 1964(1965), Long range dispersal of insects in relation to synoptic meterology. *Proc. Intern. Congr. Entomol. 12th (London)*, 6: 418–419.

French, R. A. and J. H. White, 1960, The diamond-back moth outbreak of 1958. *Plant Pathol.*, 9: 77–84.

Goodall, D. W., 1952, Quantitative aspects of plant distribution *Biol. Rev.*, 27: 194–245.

Grant, P. R., 1965, The adaptive significance of some size trends in island birds. *Evolution*, 19: 355–367.

——, 1966, Ecological compatibility of bird species on islands. *Am. Naturalist*, 100: 451–462.

Gressitt, J. L., 1954, *Insects of Micronesia. Introduction.* (B. P. Bishop Museum), 1: 1–257.

——, 1961, Problems in the zoogeography of Pacific and Antarctic insects. *Pacific Insects Monographs*, 2: 1–94.

——, 1964, Insects of Campbell Island. Summary. *Pacific Insects Monographs*, 7: 531–600.

Hamilton, T. H. and N. E. Armstrong, 1965, Environmental determination of insular variation in bird species abundance in the Gulf of Guinea. *Nature*, 207: 148–151.

Hamilton, T. H., R. H. Barth, Jr., and I. Rubinoff, 1964, The environmental control of insular variation in bird species abundance. *Proc. Natl. Acad. Sci. U.S.*, 52: 132–140.

Hamilton, T. H. and I. Rubinoff, 1967, Measurements of isolation for environmental predictions of insular variation in endemism or sympatry for the Darwin finches in the Galapagos Archipelago. *Am. Naturalist* (in press).

Horn, H., 1966, Measurement of overlap in comparative ecological studies. *Am. Naturalist*, 100: 419–424.

Hurst, G. W., 1964, Meterological aspects of the migration to Britain of *Laphygma exigua* and certain other moths on specific occasions. *Agr. Meterology*, 1: 271–281.

Hutchinson, G. E., 1959, Homage to Santa Rosalia, or Why are there so many kinds of animals? *Am. Naturalist*, 93: 145–159.

Johnson, C. G., 1957, The distribution of insects in the air and the empirical relation of density to height. *J. Animal Ecol.*, 26: 479–494.

——, 1963, The aerial migration of insects. *Sci. Am.*, 209: 132–138.

Johnson, C. G., L. R. Taylor, and T. R. E. Southwood, 1962, High altitude migration of *Oscinella frit* L. (Diptera: Chloropidae). *J. Animal Ecol.*, 31: 373–383.

Koopman, K. F., 1958, Land bridges and ecology in bat distribution on islands off the northern coast of South America. *Evolution*, 12: 429–439.

Lack, D., 1942, Ecological features of the bird faunas of British small islands. *J. Animal Ecol.*, 11: 9–36.

——, 1947, *Darwin's Finches*. Cambridge University Press.

——, 1954, *The Natural Regulation of Animal Numbers*. Oxford University Press.

Levins, R., 1966, The strategy of model building in ecology. *Am. Sci.*, 54: 421–431.

Lewontin, R. C., 1965, Selection for colonizing ability. Pp. 77–94 in H. G. Baker and G. L. Stebbins, eds., *The Genetics of Colonizing Species*. Academic Press.

Lotka, A. J., 1925, *Elements of Mathematical Biology*. Williams and Wilkins, Baltimore. (Dover reprint, 1956.)

MacArthur, R. H., 1962, Some generalized theorems of natural selection. *Proc. Natl. Acad. Sci. U.S.*, 48: 1893–1897.

MacArthur, R. H. and R. Levins, 1964, Competition, habitat selection and character displacement in a patchy environment. *Proc. Natl. Acad. Sci. U.S.*, 51: 1207–1210.

MacArthur, R. H. and E. Pianka, 1966, On optimal use of a patchy environment. *Am. Naturalist*, 100: 603–609.

MacArthur, R. H., H. Recher, and M. Cody, 1966, On the relation between habitat selection and species diversity. *Am. Naturalist*, 100: 319–327.

MacArthur, R. H. and E. O. Wilson, 1963, An equilibrium theory of insular zoogeography. *Evolution*, 17: 373–387.

Maguire, B., Jr., 1963a, The passive dispersal of small aquatic organisms and their colonization of isolated bodies of water. *Ecol. Monographs*, 33: 161–185.

——, 1963b, The exclusion of *Colpoda* (Ciliata) from superficially favorable habitats. *Ecology*, 44: 781–784.

Margalef, R., 1958, Mode of evolution of species in relation to their places in ecological succession. *Intern. Congr. Zool. 15th*, 10: paper 17.

Matthew, W. D., 1939, *Climate and Evolution*. Special Publ. New York Acad. Sci., 1: i–xii, 1–223.

Mayr, E., 1942, *Systematics and the Origin of Species*. Columbia University Press.

——, 1943, The zoogeographic position of the Hawaiian Islands. *Condor*, 45: 45–48.

——, 1951, Bearing of some biological data on geology. *Bull, Geol. Soc. Am.*, 62: 537–546.

——, 1954, Change of genetic environment and evolution. In J. Huxley, A. C. Hardy, and E. B. Ford, eds., *Evolution as a Process*. George Allen and Unwin.

——, 1963, *Animal Species and Evolution*. Belknap Press of Harvard University Press.

——, 1965a, The nature of colonizations in birds. Pp. 29–47 in H. G. Baker and G. L. Stebbins, eds., *The Genetics of Colonizing Species*. Academic Press.

——, 1965b, Avifauna: turnover in islands. *Science*, 150: 1587–1588.

Mecham, J. S., 1961, Isolating mechanisms in anuran amphibians. In W. F. Blair, ed., *Vertebrate Speciation*. University of Texas Press.

Millspaugh, C. F., 1907, Flora of the Sand Keys of Florida. *Field Columbian Museum Chicago Publ., Bot. Ser.*, 2: 191–192.

Niering, W. A., 1963, Terrestrial ecology of Kapingamarangi Atoll, Caroline Islands. *Ecol. Monographs*, 33: 131–160.

Paine, R. T., 1966, Food web complexity and species diversity. *Am. Naturalist*, 100: 65–75.

Preston, F. W., 1962, The canonical distribution of commonness and rarity: Part I. *Ecology*, 43: 185–215; Part II. *Ibid.*, 43: 410–432.

Robbins, C. S., B. Brunn, and H. S. Zim, 1966, *Birds of North America*. Golden Press.

Schoener, T. W., 1965, The evolution of bill size differences among sympatric congeneric species of birds. *Evolution,* 19: 189–213.

Sheppe, W., 1965, Island populations and gene flow in the Deer Mouse, *Peromyscus leucopus. Evolution,* 19: 480–495.

Simpson, G. G., 1950, History of the fauna of Latin America. *Am. Scientist,* 38: 361–389.

——, 1964, Species density of North American Recent mammals. *Syst. Zool.* 13: 57–73.

Slobodkin, L. B., 1961, *Growth and Regulation of Animal Populations.* Holt, Rinehart and Winston.

Small, J. K., 1933, *Manual of the Southeastern Flora.* University of North Carolina Press.

Soulé, M., 1966, Trends in the insular radiation of a lizard. *Am. Naturalist,* 100: 47–64.

Taylor, L. R., 1960, Mortality and viability of insect migrants high in the air. *Nature,* 186: 410.

——, 1965, Flight behaviour and aphid migration. *Proc. N. Central Branch Entomol. Soc. Am.,* 20: 9–19.

Van Valen, L., 1965, Morphological variation and width of ecological niche. *Am. Naturalist,* 99: 377–390.

Volterra, V., 1926, Variazione e fluttuazioni del numero d'individui in specie animali conviventi. *Mem. Accad. Nazl. Lincci,* 2: 31–113. (Abridged translation in Chapman, R. N., 1931, *Animal Ecology,* McGraw-Hill, New York.)

Watson, G., 1964, Ecology and evolution of passerine birds on the islands of the Aegean Sea. Ph.D. Thesis, Dept. of Biology, Yale University.

Wetmore, A., 1957, The birds of Isla Coiba, Panama. *Smithsonian Inst. Misc. Collection,* Vol. 134, No. 9.

Wiens, H. J., 1962, *Atoll Environment and Ecology.* Yale University Press.

Williams, C. B., 1964, *Patterns in the Balance of Nature and Related Problems in Quantitative Ecology.* Academic Press.

Willis, E. O., 1966, Interspecific competition and the foraging behavior of plain-brown woodcreepers. *Ecology,* 47: 667–672.

Wilson, E. O., 1959, Adaptive shift and dispersal in a tropical ant fauna. *Evolution,* 13: 122–144.

198

——, 1961, The nature of the taxon cycle in the Melanesian ant fauna. *Am. Naturalist*, 95: 169–193.

——, 1965, The challenge from related species. Pp. 7–27 in H. G. Baker and G. L. Stebbins, eds., *The Genetics of Colonizing Species*. Academic Press.

Wilson, E. O. and G. L. Hunt, Jr., 1967, The ant fauna of Futuna and the Wallis Islands, stepping stones to Polynesia. *Pacific Insects* (in press).

Wilson, E. O. and R. W. Taylor, 1967, An estimate of the potential evolutionary increase in species density in the Polynesian ant fauna. *Evolution* 21:1–10.

Wolfenbarger, D. O., 1946, Dispersion of small organisms. *Am. Midland Naturalist*, 35: 1–152.

Wollaston, T. V., 1877, *Coleoptera Sanctae-Helenae*. John van Voorst, London.

Zimmerman, E. C., 1948, *Insects of Hawaii. 1. Introduction*. University of Hawaii Press.

Index

adaptive radiation, 173
Aedes, 126
Aegean bird fauna, 20
Allan, H. H., 120
Antarctica, 119
ants, 10, 63, 81, 92, 117, 158, 165, 172, 178
area-species curve, 8, 172
Armstrong, N. E., 16
Asteraceae, 81
Auckland Islands, 120

Baker, H. G., 68, 163
Balanus, 98
bananaquit, 99
Barro Colorado bird fauna, 104
Barth, R. H., Jr., 19
bats, 19
beetles, 80, 85, 120, 158
birds, 16, 20, 23, 24, 43, 46, 81, 92, 97, 99, 102, 103, 104, 107, 109, 110, 115, 120, 151, 165, 173, 176
birth rate of organisms, 68
Blake, E. R., 116
Bossert, W. H., 160
Bowman, H.H.M., 53
Brown, T., 120
Brown, W. L., 159
Brunn, B., 116

Cadiz township woodland, 4
Calandra, 85
Cameron, W. A., 106
Campbell Island, 120
Carabid beetles, 158
Carlquist, S., 158, 167
character displacement, 159, 167
character release, 166
closed community, 94
clumping of islands, 29
clutch size in birds, 151
coarse-grained environments, 95, 107, 167
Cockbain, A. J., 132
Cody, M., 99, 109, 151

Coereba, 99
Coiba bird fauna, 103
Cole, L. C., 84
colonization, measurement of rate, 41; effects on evolution, 145
colonization models, 41, 51, 68
Colpoda, 82
competition, 22, 94, 107, 163, 167
Compositae, 158
compressibility, 105
compression hypothesis, 108
continental biotas, 114
convergence in evolution, 167
Cooke, C. M., Jr., 139
Crowell, K., 82, 97
Cuba bird fauna, 99
Curtis, J. T., 4
Cyanerpes, 99

Dammerman, K. W., 43
Darlington, P. J., 8, 140, 158, 175, 183
Darwin, C., 3, 158
Davis, J. H., 53
death rate of organisms, 68
demographic parameters, 68
diatoms, 15, 55
Dipsacaceae, 81
dispersal, form of curves, 124; loss of ability, 157
displacement (character), 159, 167
distance effect, 22
diversity studies in local bird faunas, 109
Dobzhansky, T., 149, 155
Docters van Leeuwen, W. M., 48
Dominica diatom flora, 15
Drosophila, 85, 155
Dry Tortugas, flora, 30, 51; ant fauna, 106
Duellman, W. E., 139
Durian (Riouw Archipelago) bird fauna, 46